Spiritual Temporalities in Late-Medieval Europe

Spiritual Temporalities in Late-Medieval Europe

Edited by

Michael Foster

CAMBRIDGE SCHOLARS PUBLISHING

Spiritual Temporalities in Late-Medieval Europe,
Edited by Michael Foster

This book first published 2010

Cambridge Scholars Publishing

12 Back Chapman Street, Newcastle upon Tyne, NE6 2XX, UK

British Library Cataloguing in Publication Data
A catalogue record for this book is available from the British Library

ISBN (10): 1-4438-2365-1, ISBN (13): 978-1-4438-2365-4

TABLE OF CONTENTS

LIST OF ILLUSTRATIONS

Acknowledgments

For one day in May, 2008, I organized a conference entitled "Belief and Time in Medieval Europe" at the Institute for Medieval Research of the University of Nottingham; the spirited and insightful studies of that conference inspired the present volume. Undoubtedly, without the institute's financial assistance and initiative, this project would not have been possible. Consideration and guidance from Rob Lutton and Judith Jesch were invaluable to me, since they were responsible for the genesis of this project; I must also thank Richard Marks and Richard Marsden for their contributions to the conference and for providing valuable feedback to the conference participants. Knowledge, wisdom, and kindness are never lacking in Thorlac Turville-Petre, to whom I shall always owe a great debt. Megan Arnott also deserves many thanks for her help. Only with her diligence and organizational skills was this project possible. Not to be forgotten are my good friends Joewon Yoon, Hikyoung Lee, and Hikyoung Moon, who supported me through the last stages of this project. Also, I am grateful for the understanding and patience of the staff at Nottingham's Institute for Medieval Research and for additional help from Tai-Won Kim and the Medieval and Early Modern Studies Association of Korea. Conversations, discussions, and friendly exchanges with Brean Hammond, Joanna Martin, Colin Gallagher, Paul Cullen, Paul Cavill, and Karen Smyth helped me in countless ways. Of course, I am most grateful to the contributors for their patience and enthusiasm at a time when medieval studies are threatened by economic disaster.

Cur venis nobiscum? Revertere et habita cum rege, quia peregrinus es et egressus de *loco tuo*
—II Samuel 15:19

Dilexi justitiam et odivi iniquitatem propterea morior in exilio
—Pope Gregory VII

INTRODUCTION

SPIRITUAL TEMPORALITIES

MICHAEL FOSTER

With *The Waning of the Middle Ages*, Johan Huizinga sparked a debate that continues to the present.[1] The premise of this book is quite simple: by the end of the middle ages, a weary pessimism had become the mood of Europe, and a sense of impending doom dominated medieval life:

> The fifteenth century, profoundly pessimistic, a prey to continual depression, could not forgo the emphatic affirmation of the beauty of life, afforded by these splendid and solemn collective rejoicings. Books were expensive, the country was unsafe, art was rare; the individual lacked the means of distraction. All literary, musical and artistic enjoyment was more or less closely connected with festivals. (230)

Huizinga's argument assumes that moments of medieval beauty are emphatic, perhaps even hysterical, attempts to escape the overwhelming sense of misery of the period. Thus his study begins and ends with a focus on the extremes of medieval emotion and life:

> All things presenting themselves to the mind in violent contrasts and impressive forms, lent a tone of excitement and of passion to everyday life and tended to produce that perpetual oscillation between despair and distracted joy, between cruelty and pious tenderness which characterize life in the Middle Ages. (2)

[1] Trans. Fritz Hopman (London, 1924). Originally published in Dutch in 1919, the English translation has been reprinted many times and remains an often-cited source for understanding the medieval worldview, although it has been largely criticized over the decades for its limited scope and selective evidence.

This hyperemotional culture is presumed not to be the norm of Huizinga's day, when a more moderate temperament makes the middle ages a seemingly foreign and miserable place:

> All this general facility of emotions, of tears and spiritual upheavals, must be borne in mind in order to conceive fully how violent and high-strung was life at that period. (6)

The misery of everyday life in the middle ages has no single source, but the conveniences and security of the modern world desensitize us to the horrors of medieval reality and encourage a nostalgic view of a simpler pastoral Elysium. Huizinga challenged historical nostalgia by reminding us that the end of the middle ages was infested with hysteria.

The Christian anticipation of an apocalyptic event implicitly encourages such a pessimistic worldview, and this attitude is confirmed by the large body of apocalyptic art and literature in the middle ages that encourages personal reform to prepare for God's final judgment. By combining an anticipation for the end of days with an introspective outlook, the medieval Christian was motivated to disregard the social, economic, and political realties around him as vanities of the flesh which would soon be irrelevant at the time of God's ultimate reckoning.

While it would be easy to dismiss all late-medieval perceptions of the flow of history to this apocalyptical anxiety, the reality of the era suggests that people went about their lives with other concerns in mind. The important collection of essays edited by John Burrow and Ian Pei demonstrates that medieval perceptions of the future were complicated and sometimes self-contradictory—indeed, much like our own.[2] A mixture of prophecy, eschatology, apocalyptical anticipation, and moral investment in the future—most tangibly in Burrow's essay on prudence as a means to gain future success in this world and the next—demonstrates a variety of attitudes to the future that often mix and are not easily divided into pragmatic, spiritual, and moralistic categories.[3]

In looking at a world as foreign as late-medieval Europe, it can be tempting to insist on how different their experiences of life and spirituality were from our own; this impulse makes each medieval text and work of art a clue of the minutiae of an exotic and unfamiliar world. The alterity of the

---

[2] *Medieval Futures: Attitudes to the Future in the Middle Ages* (Woodbridge, 2000).

[3] "The Third Eye of Prudence", *Medieval Futures*, 37-49

middle ages is obvious to any casual reader of Chaucer, yet that same reader cannot go away from the *Canterbury Tales* without some sense that we share much with our medieval ancestors: we are born; we die; we ponder our relationship to society and to the universe (however we understand its topography); we struggle to assert our own presumed correct morality on a world of competing ideologies and interests; and we try to make sense of our brief time on Earth. The variety of answers to these large questions is much greater today than it was for people in the European middle ages, yet it is difficult to find philosophical, moral, or religious ideas that have been completely abandoned as medieval naiveté. The anxiety for the apocalypse is thriving today perhaps more than ever in many parts of the United States, and the religious impulse is familiar to many modern European Union countries, where Christian Democratic political parties are a major contributor to mainstream political discourse.

Today, many feel confident that modern perceptions and understandings of time have become more complex, nuanced, sophisticated, and accurate. In the realms of science, this is undeniable; the second law of thermodynamics and post-Einstein concepts of space-time and exotic concepts such as the Planck time, relativity, and the Big Bang are substantially more consistent and independently verifiable theories of time than anything developed in the medieval or Christian traditions, yet they are either too large, too small, or too abstract to form part of our everyday experience. Even physicists count their time in minutes and hours.

Similarly, the development of the second-hand and the mechanical clock have given us tools to divide our moments into increasingly smaller units, but our lifetimes are more often divided into the larger segments of days, months and years, and many make distinctions between them by the senses. Thus sixteenth-century men and women were guided by the seasons in their daily lives and in their organization of their lives, as Amy Orrock's chapter demonstrates. The presence and experience of the seasons and their various festivals was a greater concern for people's personal and social experiences of Christianity, while the philosophical and scientific puzzles of the problem of the calendar—which would only finally be resolved with the Gregorian calendar in 1582—seem to exist in the background for most medieval Christians, just as the developments of theoretical physics are an intellectual pursuit removed from everyday life for us.

Contributors to this book attempt to assess religious understandings and experiences of time in the middle ages from different perspectives and disciplines, but all share an understanding of time as a personal and social experience beyond the quantifiable units of measure made problematic by moveable feasts. These essays encourage us to see both the similarities and differences between medieval and modern perceptions of time within the frameworks of Christian theology and everyday experience. Our focus is on instances of popular, non-canonical, or local shifts in the perception of time within a Christian world-view.

On the larger level of human history, modern perceptions of time sometimes differ from those found in the late-medieval period, since we are influenced by Kantian and Hegelian perceptions of a progressive metanarrative. Although they differ in ultimate ends, Kant and Hegel agree that human history has a distinct and definable pattern that is leading somewhere. Modern myths of progress have become the mainstays of national identities, and the totalitarian regimes of Nazi Germany, the USSR, Maoist China, and North Korea implicitly defend their control of political, economic, and social structures through the rhetoric of progress to ultimate ends.

The western world, although more multivocal, has had its own nationalist rhetoric of progression. As the USSR collapsed, confidence in America's liberal economics and representative democracy lead to musings that the end of this narrative had already arrived in the form of the socio-political reality of the United States of America.[4]

Such rhetoric of national, economic, and political progress is absent from medieval records, yet people of the era had a sense of a grand narrative leading towards an ultimate resolution. John Lance Griffith examines the differences between modern and medieval "universal histories" with the suggestion that both provide theoretical tools to understand one's individual significance in relation to one's society and to the world as a whole. The important difference is that modern universal narratives connect the individual to a project of secular improvement, leading to inevitable supermen, whereas the medieval universal narrative of St. Augustine provides meaning for the individual in his relationship to the eternal, which is fixed outside of time and unchanging. Thus every life becomes symbolic and connected to the divine, instead of one part of an evolutionary chain contributing to the good of the nation-state or the social group.

The interaction between experiencing the passing of time and maintaining Christian faith is what we refer to as "temporal spiritualities"; we suggest that the interaction between Christian faith and the passing of

---

[4] Francis Fukuyama, *The End of History and the Last Man* (New York, 1993).

time is a hermeneutic circle in which both influence each another. Thus the more time passes, the more Christian faith will change, and the more Christian faith changes, the more its believers will see time differently. The various issues studied in this book demonstrate that Christianity is a socially mediated and highly variable element of European medieval culture. Likewise, changes in Christianity will influence believers' perceptions of their lives and time itself. This can be seen quite clearly in the controversy of medieval pessimism discussed above; the presence of and faith in apocalyptic literature suggests that believers will experience an anxiety about the end of days—in fact, such literature encourages it. One need not look far in the Revelations of John to find urges for the reader to take heed of its text soon:

> et dicit mihi ne signaveris verba prophetiae libri huius tempus enim prope est
> qui nocet noceat adhuc et qui in sordibus est sordescat adhuc et iustus iustitiam faciat adhuc et sanctus sanctificetur adhuc
> ecce venio cito et merces mea mecum est reddere unicuique secundum opera sua
>
> And he saith to me: Seal not the words of the prophecy of this book: for the time is at hand.
> He that hurteth, let him hurt still: and he that is filthy, let him be filthy still: and he that is just, let him be justified still: and he that is holy, let him be sanctified still.
> Behold, I come quickly; and my reward is with me, to render to every man according to his works. (*Revelations* 22:10-12)[5]

It is still unclear to what extent the apocalypse was anticipated in the late middle ages; judging by the presence of Middle English poetry on the subject, it seems to have been a popular component of Christian faith, yet the abundant attempts in medieval law to control the status of assets well into the future, as Paul Brand has discussed,[6] suggests the presence of a more pragmatic belief that the world would be around for a while, and thus it needed to be controlled through social conventions.

Legal and philosophical debates share a practical concern for people's relationships to posterity, and assume that a human civilization will

[5] The translation is from the Douay-Rheims Bible.

[6] Paul Brand, *In perpetuum: The Rhetoric and Reality of Attempts to Control the future in the English Medieval Common Law*, in *Medieval Futures* 101-113.

continue to mediate reputation and social standing. Laura Cleaver's chapter in the present volume examines this phenomenon in relation to Peter Lombard's contested status throughout the twelfth and early-thirteenth centuries; her essay suggests that artists and writers wanted to contribute to an academic debate on the thinker's status that was expected to continue on Earth amongst living human beings.

This lengthy debate was an attempt to construct meaning and significance in relation to Peter Lombard, and many medieval authors and artists were more invested in the long-term project of constructing and mediating meaning than they were in self-expression or self-promotion. Similarly, my essay on scribal variants in *Tundale* argues that scribes in the middle ages were often motivated by an urge to correct readers' understandings of texts, and they consciously edited their texts as they responded to them. Such rhetorical emendation suggests that medieval doctrine was not fixed dogma, but a malleable collection of ideas and theories that were subject to variation and change as individual Christians modified and emended the tradition they inherited. Importantly, this individual "improvement" does not come from centers of authority, and it is not explicitly sanctioned by the Church.

A popular understanding of the relationship between real, contemporary time and Biblical time can be found in the Middle English York cycle plays, which dramatically realize Biblical events not as symbolic or allegorical concepts, but as events occurring on a real timeline. At the same time, these plays incorporate contemporary concerns and interests with the ancient, Biblical content. Karen Ward analyzes the rhetorical effect of the balance between elements of contemporary reality and representations of Biblical history within the York plays; her reading of *The Crucifixion* illuminates a popular, communal understanding of the similarities and differences between Biblical and contemporary time in late fourteenth-century England while it emphasizes the central importance of Christian salvation to their own lives.

The impulse of bookmakers to modify texts to new ideological ends is also the focus of Nancy Ross's chapter on images accompanying the commentary on the apocalypse myth by Berengaudus of Ferrièrs in two English manuscripts of the late thirteenth century. Ross demonstrates how the manuscripts present an anti-Semitic argument that focuses on the presumed contemporary sins of the Jews in medieval Europe while ignoring those "ancient crimes" that led to the crucifixion of Christ. Here we see a combination of image and text organized to form a political argument that culminated in the expulsion of the Jews from England in 1290. The presentation of this commentary in these books transforms ancient beliefs in an attempt to modify medieval attitudes while maintaining and even intensifying old prejudices and hatred.

Luciana Meinking Guimarães finds much larger shifts in motifs and themes amongst the Welsh saints' lives, which form a cohesive corpus of texts that are often seen as propaganda-responses to the Norman invasion and occupation of Wales. Guimarães finds instead a collection of motifs that suggest the perpetuation of a Welsh hagiographical tradition that precedes the extant manuscripts. At the same time, these themes are adapted to a new historical context where they take on new significance. Here, as in *Tundale* and the Berengaudus commentary, we see a popular Christian tradition being perpetuated and adapted to a new moment in time.

The adaption of old works of religious art to a new secular situation is also the concern of Tom Nickson's chapter, which suggests that the changing collection of images in Toledo Cathedral reflects a philosophical response to Gregory's defense of the use of images in articulating belief. Thus the ornamentation of the building itself becomes a battleground for a burgeoning theological debate, and artists are invited to take part in the debate with their contributions to the building's artworks. Here we see archbishop Tenorio, a central and powerful figure of the Church, asserting the papal defense of religious imagery by reshaping a part of Christendom's topography.

We hope that these essays will contribute to an understanding of medieval perceptions and experiences of time not as a completely alien historical phenomenon, nor as entirely consistent with modern perceptions of the world. Rather, we would like to introduce these instances of spiritual temporalities with the hope that further comparative analysis will uncover how the myths and realities of modern life mirror the religious perceptions of and responses to the passing of time in late-medieval Europe.

Bibliography

Brand, Paul. "*In Perpetuum*: The Rhetoric and Reality of Attempts to Control the Future in the English Medieval Common Law", in *Medieval Futures: Attitudes to the Future in the Middle Ages* (Woodbridge, 2000), 101-113.

Burrow, J. A. and Ian Pei, eds. *Medieval Futures: Attitudes to the Future in the Middle Ages* (Woodbridge, 2000).

Burrow, J. A. "The Third Eye of Prudence", in *Medieval Futures: Attitudes to the Future in the Middle Ages* (Woodbridge, 2000), 37-49.

Fukuyama, Francis. *The End of History and the Last Man* (New York, 1993).
Huizinga, Johan. *The Waning of the Middle Ages*, trans. Fritz Hopman (London, 1924).

CHAPTER ONE

AUGUSTINE AT THE END OF HISTORY: UNIVERSAL HISTORIES AND THE INDIVIDUAL'S PLACE IN TIME

JOHN LANCE GRIFFITH

In his *Life of King Alfred*, Asser recounts the story of Alfred's attempt to overcome a problem of time. The king had promised to devote "one half of his mental and bodily effort both by day and by night" to the service of God,[1] yet he found that he "could not in any way accurately estimate the duration of the night hours because of darkness, nor the day-time because of the frequent density of rain and cloud".[2] And so he invents a time-keeping device, a series of wax candles that burn at regular intervals continuously throughout the day.

We are reminded that, at a very basic level, the medieval experience of time was quite different from our own. We have constructed social and cultural systems which simply cannot function without precise dividing, measuring, and tracking of time. But even after the invention of accurate time pieces, how differently time must have been filled by the medieval person--how differently time must have been perceived by the medieval mind. As Augustine observed, long before Einstein, we "must not allow [our] mind to insist that time is something objective... I measure time in my mind. For everything which happens leaves an impression on it, and this impression remains after the thing itself has ceased to be. It is the

[1] Alfred the Great, *Alfred the Great: Asser's Life of King Alfred and Other Contemporary Sources*, trans. Simon Keynes and Michael Lapidge (New York, 1983), 107.
[2] Ibid.,108.

impression that I measure."[3] And those medieval impressions must be in many ways strange, if not unknowable, to us who move at speeds and process relentless streams of data from omnipresent television and internet media unthinkable in the medieval world.

Yet, however differently the medieval and the modern mind may perceive the speed of passing time when observing the other—however "fast" this world might seem to the time traveler from the middle ages, however "slow" the daily grind of the medieval worker to us—surely the reflecting on one's own time in the world leads one (medieval or modern) to the conclusion that it is all too brief; that our allotted interval (as Walter Pater would say) is marked by an awful brevity. Chaucer's Host observes:

> Lordynges…I warne yow, al this route,
> The fourthe party of this day is gon.
> Now for the love of God and of Seint John,
> Leseth no tyme, as ferforth as ye may.
> Lordynges, the tyme wasteth nyght and day,
> And steleth from us, what pryvely slepynge,
> And what thurgh necligence in oure wakynge,
> As dooth the streem, that turneth nevere agayn,
> Descendynge fro the montaigne into playn.
> Wel kan Senec and many a philosophre
> Biwaillen tyme, moore than gold in cofre;
> For "Los of catel may recovered be,
> But los of tyme shendeth us," quod he.
> It wol nat come agayn, withouten drede,
> Namoore than wole Malkynes maydenhede,
> Whan she hath lost it in hir wantownesse.
> Lat us nat mowlen thus in ydelnesse.[4]

Time is a thief, a ravager, a wasting force, as in the Reeve's metaphor of life as a draining barrel of ale:

> As many a yeer as it is passed henne
> Syn that my tappe of lif bigan to renne.
> For sikerly, whan I was bore, anon
> Deeth drough the tappe of lyf, and leet it gon,
> And ever sithe hath so the tappe yronne

[3] Augustine, *Confessions*, trans. R.S. Pine-Coffin (New York, 1961), 276.

[4] *Introduction to the Man of Law's Tale*, ll.16-32. All references to Chaucer are from Geoffrey Chaucer, *The Riverside Chaucer*, 3rd edition, ed. Larry D. Benson (Boston, 1987).

Til that almoost al empty is the tonne.
The streem of lyf now droppeth on the chymbe.
The sely tonge may wel rynge and chymbe
Of wrecchednesse that passed is ful yore;
With olde folk, save dotage, is namoore![5]

For the Host, the way to battle time is to try and keep ahead, finding the pleasure in a good tale while one can. For the Reeve, the battle for the body is lost, though there remains a spark of spirit, and emotions (even negative ones) keep one struggling on, finding, as he says, a ripeness in the rottenness.[6] For both, the implication is that life is short and death (the end of time) is always a nearby possibility. It is partly this sense of brevity and of the unpredictable arrival of death that makes the Host and the other pilgrims both likeable and intelligible to the modern reader. Our shared perception of the wasting and our fear of death transcends the historical and cultural time that separates us.

It is then not so much our respective conceptions of time that marks a significant division between the medieval and the modern, as much as our respective ideas of history—that is, understandings of the purpose of time and of our place (be it as individuals or communities and generations) in time. This essay examines the concept of universal history, the way in which that concept joins and divides the medieval and the modern imagination. I refer to histories of the world from beginning to end, theories of time which posit an unfolding of time towards an intelligible end—the world-historical views of thinkers like Augustine, Kant, Hegel, and Marx.

Initially a response to pagan critics who blamed the spread of Christian values among Roman citizens for the sack of the city, Augustine's *City of God* eventually became an analysis not just of current Roman events or even of the history of the Empire, but an explanation for all of human history—that is, a universal history. Though he does not invoke the phrase, Augustine is a seminal figure in the history of the idea of "universal history." His opus is one of the most influential outlines of universal history ever written, shaping the medieval world's understanding of history and time, and of the individual's place in both. More modern universal histories such as those of Kant, Hegel, and Marx, vary widely from one another and of course deviate substantially from Augustine's medieval understanding. But all such histories have the same basic

[5] *The Reeve's Prologue*, ll.3889-3998.
[6] Ibid., l.3875 ("Til we be rotten, kan we nat be rype").

purpose: to establish the meaning of human existence through an investigation of the teleology of its history, to understand the relevance of one's own time through its relation to the beginning and ending of a course of events unfolding in time. So while one way to mark historical periods is by changes in theories of universal history (marking Kant's Enlightenment philosophy from Augustine's Medievalism), the break between the periods with and without such a concept of history is also significant. For those in the modern world without a sense of teleological history (or who believe that they are living at what Hegel would term "the end of history"), Augustine's history highlights a radical difference between the medieval and the modern mind.

Moreover, exploring Augustine's place in the history of the idea of universal history helps us to consider how such a conception of time and history impacted literary narratives such as those of Chaucer in the Middle Ages; that is, how this view of history—which may seem puzzling to those of us who have arrived at the end of history—affected their understanding of the place of the individual, the universal human pilgrim, as it were, in time.

A Modern View of Universal History

What does it mean for the modern individual to be part of a universal history or to be living at the end of history? Where Augustine imagines a literal beginning and end of the world (where time and existence as we know it began and will cease to be), the post-Enlightenment universal histories from Kant through Marx tend to concentrate on the end point which is not apocalyptic but liminal in the sense that significant cultural/political change is no longer possible or desirable. Comprehending the vast scope of time needed for such progress requires political and economic philosophy, not theology. "History," the unfolding of events and the record of change wrought by political and economic forces, ends—for Kant in the establishment of a global union of states, for Marx in a socialist utopia.

How one evaluates this moment in history at the end of history depends in part on what one takes to be evidence of "significant" change; and also on whether one emphasizes the "us" or the "me" implicit in this evaluation, whether one emphasizes the consequences for the collective culture or for the individual.

The history of the individual is like and unlike the history of the world. For Alexandre Kojève (whose lectures on Hegel early in the twentieth century crystallized the Hegelian concept of universal history for subsequent generations of thinkers), the two are directly linked, since the individual, like history, is defined by movement through time, not space. "Man becomes conscious of himself at the moment when—for the 'first time'—he says 'I.'"[7] That "I" is revealed through its desire (not an animal desire for a natural object easily satisfied, but desire for desire itself) and is revealed in its pursuit of what it is not; that is, revealed in its "becoming" in time:

> This I, which "feeds" on Desires, will itself be Desire in its very being, created in and by the satisfaction of its Desire. And since Desire is realized as action negating the given, the very being of this I will be action. This I will not, like the animal "I," be "identity" or equality to itself, but "negating-negativity." In other words, the very being of this I will be becoming, and the universal form of this being will not be space, but time.[8]

To talk meaningfully about the individual is to talk not about what he is, but what he is becoming. In this case, the end toward which his own dissatisfaction is driving him to evolve. Self-consciousness is a process, a production subject to constant change and, like history, unfolded continuously in time rather than defined fixedly in space:

> Therefore, its continuation in existence will signify this I: "not to be what it is (as static and given being, as natural being, as 'innate character') and to be (that is, to become) what it is not." Thus, this I will be its own product: will be (in the future) what it has become by negation (in the present) of what it was (in the past), this negation being accomplished with a view to what it will become. In its very being this I is intentional becoming, deliberate evolution, conscious and voluntary progress; it is the act of transcending the given that is given to it and that it itself is. This I is...historical (in relation to itself).[9]

But how does this "historical" individual relate to history itself? In Kojève's reading of Hegel, the evolution of individuals and the progress of their individual desires plays out as a struggle between masters and slaves. This struggle *is* history, a process of becoming, of cultural identity established in time rather than space, similar to that of the individual. This is in part because, for Kojève, desire here is specifically the human desire for desire (as opposed, we said, to a simpler animal desire for mere

---

[7] Alexandre Kojève, *Introduction to the Reading of Hegel* (Ithaca, 1969), 3.
[8] Ibid., 5.
[9] Ibid., 5.

material objects which please us); substituting the self for the traditional object of desire, human desire becomes a desire for recognition of the individual's own self-worth. This inevitably leads to a conflict between individuals, as they struggle not for desired resources, but for recognition. Each tries "to impose itself on the other as the supreme value; accordingly their meeting can only be a fight to the death."[10] The desire for recognition becomes a desire for power to ensure such recognition of self worth. History is the record of this struggle, between the masters who use their power to impose values and the slaves who resist or, if unwilling to risk their lives in the struggle, acquiesce:

> If the human being is begotten only in and by the fight that ends in the relation between Master and Slave, the progressive realization and revelation of this being can themselves be effected only in terms of this fundamental social relation. If man is nothing but his becoming, if his human existence in space is his existence in time or as time, if the revealed human reality is nothing but universal history, that history must be the history of the interaction between Mastery and Slavery: the historical "dialectic" is the "dialectic" of Master and Slave.[11]

If the urge for self-recognition, and the desire for the power to ensure such recognition, is so basic to the human condition, how is it possible for such a process to come to an end? How can an existence defined by its change and struggle in time come to an end except through an apocalyptic destruction of time itself? Hegelian logic drives Kojève to the inevitable answer:

> But if the opposition of "thesis" and "antithesis" is meaningful only in the context of their reconciliation by "synthesis," if history (in the full sense of the word) necessarily has a final term, if man who becomes must culminate in man who has become, if Desire must end in satisfaction, if the science of man must possess the quality of a definitively and universally valid truth—the interaction of Master and Slave must finally end in the "dialectical overcoming" of both of them.[12]

History, as a dialectical process, contains within itself the solution, the processes by which its own nature will be overcome, by which it will be transformed into its final state.

---

[10] Ibid., 7.
[11] Ibid., 9.
[12] Ibid., 9.

Masters and servants are not just byproducts of human history, but essential elements which will bring about the end of history: individuals struggle for recognition and power; some become masters and some (unwilling to risk death) become slaves; but the slaves then struggle against the masters, using their own work as a means of establishing their self-worth and so eliminating their dependence on masters; the overcoming of the masters through work puts an end to the history of struggle, and an end to history itself.[13] But the essential point is that slaves require masters to force them to work, to develop the concept of work which will ultimately free them. History, then, may or may not have been "designed" or "planned" by a rational being, but it is governed by a logic that compels the development of the individual (the development of modern self-consciousness) and of human society to a particular end. Because this process is a temporal one, time (more so than space) is essential to the modern individual's understanding of himself and of the modern world.

Recalling the Host's image of time as a stream descending from a mountain to a plain and the Reeve's image of time as a slowly emptying barrel, we note that in both cases there is no indication of progress. There is structure, order, even logic, but not the sense that conceives of movement to a better place (in fact, the Host's image is one of descent), or of the radical change for the better implicit in most modern conceptions of universal history (the Reeve's cask, in going nowhere, returns only to its original state of emptiness).

Both are images of an individual's relation to time, of time allotted specifically to the individual, though they could as well be figures of time and of human history itself, if both the Host and the Reeve did not lack a sensitivity to the theology that makes possible the Augustinian concept of history and were not seemingly unaware of the individual's relation to divine history. From an Augustinian point of view, one could say that the Host and the Reeve have missed the point (or misread the signs): the stream having descended the mountain is purified, calmed after its rough descent, and its origin (which the Host laments cannot be returned to) is insignificant given that now it placidly flows on without end; similarly, the Reeve's empty cask can be seen as a purification of the soul, emptied of desire and turbulent passion, the individual returned to the nothingness, the blank surpassing human imagination and understanding, which is God

[13] Ibid., 24-27.

(the "unknown Nothingness"/ *nihil incognitum* as Bataille says, quoting Saint Angèle de Foligno).[14]

Still, even in the Augustinian view, the radical change (the progress) of the individual or of the society does not happen in time or in history, but rather at the end of time (at the moment of the individual's death or at the apocalypse, the end of all society). Or rather, the degree to which the change is perceptible seems much less in the medieval than in the modern view. Of course, as discussed below, Augustine does imagine the individual participating in divine history, contributing to the end of history, since the City of Man and the City of God exist simultaneously and are interconnected. And we should not overstate the static nature of medieval culture; it could be dynamic and there was more social mobility than clichéd images of the period suggest. Yet in contrast to the modern era, medieval culture and the individual's movement within that culture are relatively static. Even in the fourteenth century, a period of greater social mobility, Chaucer's pilgrims are defined for the most part not by names but by occupation or type. Times change, but not by very much; individuals come and go, but their social roles are basically the same. Though there is free will and room for the individual to learn and the soul to develop and to make better choices, medieval discourse allows no space for attaining "self-consciousness," which Kojève defines as the continuous and conscious development of self through "becoming." The classical world could be easily fused (or confused) with the medieval world by Chaucer's Knight because the historical moment beyond which the ancients reside is the Incarnation; what antiquity lacked was knowledge of Christianity, and so the "progress" of the fourteenth century was somewhat narrow in scope, at least in comparison to what the Enlightenment would later conceive of as historical progress. Medieval history is moving toward an end, a better world, and the individual has a role to play (perhaps, as we will see, more so than in modern conceptions of universal history), but this is not the linear progressive history of the Enlightenment.

From a Hegelian point of view, the individual and the world both pass from history into the end of history; for Hegel and Kojève, from a political and cultural perspective, this end has already arrived (or very nearly arrived) and is for the best: slaves have freed themselves from their masters and, upon the principles of liberty and equality, a union of people

---

[14] Georges Bataille, *Inner Experience*, trans. Leslie Anne Boldt (Albany, NY, 1988), 104.

and of states has emerged to provide relative peace and stability. In Francis Fukuyama's terms, liberal democracies like the American federal system and the European Union have emerged.[15] Although such systems are always in need of some small correction, they are the final form of political organization in that their values (of liberty and equality) are the most universal and most likely to become globalized. Human society, the world, has achieved its end and will persist in this state. History, as the record of significant change, has ended.

The concept of universal history has produced, of course, its share of critics; aside from inviting debates about what the end of history will actually be like (Kojève's and Fukuyama's vision as opposed to Marx's), the idea of history being an irresistible social force leading us to an inevitable outcome is simply unacceptable to many modern thinkers. Karl Popper's main objection to Hegel was that the positing of a historical dialectic diminishes the human capacity for free will; if the end of history is certain, then we are in no position to resist or change that end, even if we do not like it:

> If we think that history progresses, or that we are bound to progress, then we commit the same mistake as those who believe that history has a meaning that can be discovered in it and need not be given to it. For to progress is to move towards some kind of end, towards an end which exists for us as human beings. 'History' cannot do that; only we, the human individuals, can do it; we can do it by defending and strengthening those democratic institutions upon which freedom, and with it progress, depends...Instead of posing as prophets we must become the makers of our fate.[16]

Notable here is that even Popper has a concept of "progress," even if he deplores the concept of history and historical forces (the determinism of which he may be somewhat overstating). But my goal here is not to choose between such views; instead, I would like to focus on not so much the political or economic implications of these ideas, but rather on what these interests reveal about modern attitudes toward time and history. Although the discourse is quite modern ("self-consciousness," "progress," "dialectic"), the fundamental philosophical problem behind these debates

---

[15] Francis Fukuyama, *The End of History and the Last Man* (New York, 1993, reprinted 2006).

[16] Karl Popper, *The Open Society and its Enemies Vol. 2: Hegel and Marx* (Princeton, 1966), 279-280.

would not be unintelligible to the medieval thinker (though, as we will see below in the discussion of Augustine, the medieval framing and solution of the problem is quite different): what is the ultimate purpose of a temporally constructed world, of a universe existing in time? And what is the purpose of the individual in the history of that universe?

Let us assume for the moment that the Hegelian end of history is good for the world society and, in the view of Kojève, good for the individuals living at the end of history, since they experience freedom and equality. Yet the individual is not the world. World history may come to a neat conclusion after the relentless unfolding of events in time; the individual, however, both experiences and reflects upon that history—he asks, what is my role in that unfolding plan, what am I to do at the end of that history? If one is not Napoleon, a Hegelian superman as it were, then what is one's purpose other than to watch these grand events unfold? And at the end, what is anyone, superman or commoner, to do, to think, to make theories about? If as a culture, a civilization, and a species we have exhausted our collective imaginations, does any possibility of significant progress present itself to the individual?

For Georges Bataille, our position at the Hegelian end of history is at once tragic and darkly comic. We face the possibility that, historically speaking, if we take history to be the record of significant change through time, there is nothing ahead of us, only a stale analysis of how we have come to be here; only distracting, if pointless, speculation about whether it was all worth it. Nothing left, Bataille says, but an "unemployed negativity," the proving of one's own "irrevocable insignificance." Bataille embraces this hypothesis "seriously though cheerfully":[17]

> But the horror he [the man at the end of history] feels looking at negativity within himself is no less likely to end in satisfaction than in the case of a work of art (not to mention religion). For it is precisely in needing to act that he has recognized negativity; and this recognition is bound up with a conception that has it be the condition of all human existence. Far from stopping in this investigation, he finds a total satisfaction in the fact of becoming the man of "recognized negativity." He will no longer rest as he begins the effort to pursue this recognition to its very end...Thus [bringing] into play representations extremely charged with emotive value (such as physical destruction or erotic obscenity, an object of laughter, of physical excitation, of fear and of tears). But at the same time these representations

[17] Georges Bataille, *The College of Sociology 1937-1939*, ed. Denis Hollier (Minneapolis, 1988), 90.

> intoxicate him, he strips off the straitjacket that has kept them from contemplation and he sets them objectively within the eruption of time that nothing changes. He understands then that it is his good, not his bad, luck that brought him into a world where there was nothing left to do. For he cannot be the man of "recognized negativity" except to the extent that he makes himself be recognized as such. Thus, once again, he discovers something "to do" in a world where, from the point of view of actions, nothing is done any more. And what he has "to do" is to satisfy the portion of existence that is freed from doing: It is all about using free time.[18]

For Bataille's weary individual employed only in establishing his own "negativity," the event horizon which marks that passing into the end of history is not unlike that moment when one realizes that one has already read Shakespeare for the first time. Whatever lies beyond that—re-readings, new readings, comparative readings, writings—the radically transformative, the radically developmental part of one's history is largely past, gone with one's youth.

Of course, with the Wordsworthian voice of the "Intimations of Immortality" ode, one could say that this is a good thing, that not all is lost "in years that bring a philosophic mind"; one can get on with the business of being an adult, a scholar, cataloguing one's memories and systematizing our knowledge of the world, even if we are no longer experiencing it in a vital way. For Walter Pater it is good as well, since the social stability derived from freedom and equality give one the opportunity to focus on the awful brevity of one's interval, to pursue the aesthetic experiences which alone bestow solace if not significance on that interval—"on this short day of frost and sun...we shall hardly have time to make theories about the things we see and touch."[19]

For Bataille, it is an ambiguous good at best; in some senses liberating, but the indulging of desire and the thrill of excess and of pushing limits ultimately proves only a negative proposition, establishing the truth of the individual's own negativity:

> I enter into a dead end. There all possibilities are exhausted; the "possible" slips away and the impossible prevails. To face the impossible—exorbitant, indubitable—when nothing is possible any longer is in my eyes to have an experience of the divine; it is analogous to a torment.[20]

---

18 Ibid., 91-92.

19 Walter Pater, *The Renaissance*, ed. Adam Phillips (Oxford, 1986), 152.

20 Bataille, *Inner Experience*, 33.

> Supposing that I were to be God, that I were to have in the world the assurance of Hegel (suppressing shadow and doubt)—knowing everything and even why fulfilled knowledge required that man, the innumerable particularities of *selves*, and history produce themselves—at precisely that moment, the question is formulated which allows human, divine existence to enter...the deepest foray into darkness without return; why must there be *what I know*? Why is it a necessity? In this question is hidden—it doesn't appear at first—an extreme rupture, so deep that only the silence of ecstasy answers it.[21]

One can understand Bataille's fascination with medieval mystics such as Saint Angèle de Foligno, who located God at the extremes of human perception and understanding, in the Nothingness beyond human experience and knowledge.[22] In accepting the reality of the end of history, there is always for Bataille a desire to resist because it is terrifying to face the possibility that life (collective human life and that of every individual) is what Chaucer's Reeve thought it was: in the end, just an empty cask that has been slowly depleted to reveal its essential (and rather unimaginative and relatively uninteresting) shape, its fundamental hollowness, its natural function to frame an empty space.

For the Reeve, as for Bataille, there is no transformation—at the beginning and at the end the Reeve is a reeve, and his work brings him no freedom, no illumination; the cask of life, the shape and the content of time, is known and found wanting. There is only the pursuit of desire, the experience of the passions (anger, jealousy, lust, etc.), and then at the end of life the chasing down of the last sparks of those passions:

> But ik am oold; me list no pley for age;
> Gras tyme is doon; my fodder is now forage;
> This white top writeth myne olde yeris;
> Myn herte is also mowled as myne heris,
> But if I fare as dooth an open-ers -
> That ilke fruyt is ever lenger the wers,
> Til it be roten in mullok or in stree.
> We olde men, I drede, so fare we:
> Til we be rotten, kan we nat be rype;
> We hoppen alwey whil that the world wol pype.
> For in oure wyl ther stiketh evere a nayl,

21 Ibid., 109.

22 Ibid., 103-104.

To have an hoor heed and a grene tayl,
As hath a leek; for thogh oure myght be goon,
Oure wyl desireth folie evere in oon.
For whan we may nat doon, than wol we speke;
Yet in oure asshen olde is fyr yreke.
 Foure gleedes han we whiche I shal devyse -
Avauntyng, liyng, anger, coveitise;
Thise foure sparkles longen unto eelde.
Oure olde lemes mowe wel been unweelde,
But wyl ne shal nat faillen, that is sooth.
And yet ik have alwey a coltes tooth,
As many a yeer as it is passed henne
Syn that my tappe of lif bigan to renne....[23]

Of Chaucer's pilgrims, the Pardoner may be the most like Bataille, the most modern; not necessarily in terms of his manipulation of and indifference to his fellow pilgrims, but in his resistance to the "system" (as Bataille referred to Hegel's philosophy), to the universe he lives in, to what he has been told it is and must be:

For lewed peple loven tales olde;
Swiche thynges kan they wel reporte and holde.
What, trowe ye, that whiles I may preche,
And wynne gold and silver for I teche,
That I wol lyve in poverte wilfully?
....
I wol nat do no labour with myne handes,
Ne make baskettes and lyve therby....[24]

The Pardoner is (as he explicitly says) motivated by greed. But in his rejection of honest work we see too his rejection of the whole history of man since the Fall (of the "tales olde" which he preaches) and of the God responsible for such a universe. He wants no part of a fallen world of labor, even though the work and the struggle of that world will ultimately lead (in a somewhat Hegelian fashion) to its salvation. Without denying the history into which he has been born, the unfolding of the universal plan from the Fall through the Incarnation to the Apocalypse, or even denying the existence of the God responsible for such a history, he simply rejects that order, turning instead to blasphemy, irony, and negation. He strangely

[23] *The Reeve's Prologue*, ll.3867-3890.
[24] *The Pardoner's Prologue*, ll.437-445.

seems to prefer (and in this he is unlike Bataille) a world without history or one at the end of history, without any God, without any work to do. The empty universe, the escape from history and from God's plan which the Pardoner longs for, is the world that Bataille, having been born into at the end of history, despairs of. In this sense, the Pardoner is foolish from both the medieval and the modern perspective. He is not a modern character and it would take a modern philosopher to find a strain of heroism in his opposition; from a medieval perspective, the Pardoner is a caricature, an extreme depiction of vice. Could anyone who believed in hell choose damnation as easily as he seems to? But his comments about labor reflect a medieval understanding of and struggle with the implications of being born into God's history, into a time-bound world which unfolds into a universal plan.

Before considering Augustine's response to such problems, let us briefly consider Kant, whose point of view suggests that the history of the individual and the history of the world are (as in Hegel and Kojève) similar, but for a different reason. Meaningful experience is not exhausted in Kant; rather, it overloads the capacity of the individual. It is not that there is nothing more (i.e., of great significance) to experience or to discover, but rather, such possibilities are beyond the power of most individuals to discover in their lifetime. No one lives long enough to experience such significant change. That is the reason Kant feels he must place his hope in his species, in social evolution, and in the future: "It remains strange that the earlier generations appear to carry through their toilsome labor only for the sake of the latter."[25] It is strange, but that is the way it works: only at the end of history is the plan revealed; in the meantime, each civilization dies with an incomplete view of this progressive history, each individual dies without experiencing the fruits of this great universal plan, for the most part living lives that seem to have no knowledge of this unfolding of rational history: "everything in the large woven together from folly, childish vanity, even from childish malice and destructiveness."[26] For Kant, most people are still children; and yet, while that means a highly chaotic, often maliciously destructive world, it also means that the best is yet to come: the transformative moment, a progress into a better world, as yet unknown.

---

[25] Immanuel Kant, *Idea for a Universal History from a Cosmopolitan Point of View*, trans. Lewis White Beck (1963), Third Thesis.
http://www.marxists.org/reference/subject/ethics/kant/universal-history.htm.
[26] Ibid., *Introduction*.

What is striking about Kant's essay is that it develops out of a sour perception of "the idiotic course of things human"[27] and yet wills itself toward a bold claim of faith in the future, in the deep structure of the universe. "Nature" for Kant must be rational; human beings cannot be relied upon to provide a plan and purpose, and so the philosopher has no other choice but to look to Nature: "no other expedient for him except to try to see if he can discover a natural purpose in this idiotic course."[28] This kind of logic exposes some of Kant's own doubts and perhaps undermines his main argument. (Nature has a rational plan—it must; because if it does not, Kant implies, we must face the possibility that life has no meaning, which we are not willing to do; and so, courting circular logic, we insist that there is a rational plan…). As Bataille found in the theories of Hegel and Kojève, one might find oneself arriving at a point of unrest, unsettled by a lingering sense of dissatisfaction with Kant's concept of universal history. It is such unrest that Augustine's much earlier universal history was designed to cure.

Augustine's *City of God*

Let us turn back to the medieval period on the timeline of universal history to the critical point in the fifth century when Augustine wrote the *City of God*.[29] This view of universal history also highlights the difference

---

[27] Ibid.

[28] Ibid.

[29] Augustine is of course not the only medieval thinker to reflect on the nature of time or of history. His reflections in Book XI of the *Confessions* on the nature of time and God's relation to time were developed and expanded upon throughout the late medieval period, especially by thirteenth and fourteenth-century scholastics. See Rory Fox, *Time and Eternity in Mid-Thirteenth-Century Thought* (Oxford, 2006). On medieval and modern conceptions of history, see D. Vance Smith, "Irregular Histories: Forgetting Ourselves," *New Literary History* 28 (1997), 161-184; Richard Landes, "The Fear of an Apocalyptic Year 1000: Augustinian Historiography, Medieval and Modern," *Speculum* 75 (2000), 97-145; and Philipp W. Rosemann, "Heidegger's Transcendental History," *Journal of the History of Philosophy* 40 (2002), 501-523. *City of God* is probably the most comprehensive and influential universal history (and so most relevant for the purposes of this essay), but for the history of other medieval conceptions of history and historicity, see Sharon Dale, Alison Willams Lewin, and Duane J. Osheim, eds, *Chronicling History: Chroniclers and Historians in Medieval and Renaissance Italy* (University Park, PA, 2007); Gabrielle M. Spiegel, *The Past as Text : The Theory and Practice*

between the history of the individual and the history of the world. For Augustine, the City of God exists in eternity; the City of Man changes, moves through time, but never achieves (as Kant hopes) the Eternal City on Earth in history—that is, in time. Unlike Kant's universal history, Augustine's history is neither evolutionary and mechanistic, nor idealized (except in so far as the Final Judgment, through the power of God, brings about radical change and ends history):

> Both of these cities [the City of Man and the City of God] alike make use of temporal goods and both are equally afflicted by temporal ills—but how different they are in faith, how dissimilar in hope, how unlike in love! This will go on until they are to be separated in the Last Judgment, when each shall achieve its appointed end—an end which will have no end.[30]

While history, the record of human existence in time, culminates in the full realization of the City of God, that event takes place outside of time. More importantly, that City is not the product of historical change. Because, like God, it is eternal and exists outside of time, there is a sense in which it is always present in the temporal span of history.

The metaphysics of Augustine's two cities, the complicated ways in which they not only co-exist but are used by Augustine to represent symbolically and contain a myriad of Augustine's ideas about good and evil, are quite interesting,[31] but they are not my particular interest here. Instead, I would like to focus on what this concept of history means for the average individual, as opposed to how universal histories serve the modern philosopher. For Kant and Hegel, the end of history is a future happening that the (average) individual never fully understands or at best experiences after it has happened. History is a process unfolding around him. The process may be influenced by a few great men, and it may be analyzed by the philosophers, but for the most part such history dwarfs the power and the significance of the individual. Theories of universal history provide an end to be imagined and a purpose for history itself, but they do not

---

of Medieval Historiography (Baltimore, 1997); Ernst Breisach, *Historiography: Ancient, Medieval, and Modern* (Chicago, 2007).

[30] Augustine, *City of God*, ed. Vernon J. Bourke with introduction by Etienne Gilson (New York, 1958), XVIII.liv.

[31] For an analysis of *City of God*, see P.A. Thurlow, "Augustine's *City of God*, Pagan History and the Unity of the Annolied," *Reading Medieval Studies* 6 (1980), 44-67; and R.A. Markus, *Saeculum: History and Society in the Theology of St. Augustine* (Cambridge, 1970).

necessarily explain the purpose of any random individual or the reason that he or she exists and suffers in time.

In Augustine, however, the individual has a more significant place and a more intimate connection to history. Although God, outside of time, necessarily sees all and knows the outcome of this entire history, for human beings in time "man's will...is all important"[32] and can be directed rightly or wrongly:

> One city is that of men who live according to the flesh. The other is of men who live according to the spirit. Each of them chooses its own kind of peace and, when they attain what they desire, each lives in the peace of its own choosing.[33]

This is to say that the individual is not only connected to the City of God, but to a certain extent contributes to it by becoming a citizen or not, by transforming the earthly time-bound world into an image of that eternal City which exists as a shadow to the historical world of man before it manifests fully outside of time at the end of history:

> A shadow, as it were, of this eternal City has been cast on earth, a prophetic representation of something to come rather than a real presentation in time. Yet this shadow, merely symbolic as it is and not the reality that is to be, is properly called the holy City...[In the biblical story of Abraham's sons, Ismael and Isaac] we see that one portion of the world community [Sara and Isaac] became a symbol of the heavenly City...It was founded, not for its own sake, but as the shadow of another substance, a shadow that was itself foreshadowed by a previous symbol [Agar and Ismael]....[34] As for the city of this world, it is neither to last forever nor even to be a city, once the final doom of pain is upon it. Nevertheless, while history lasts, it has a finality of its own; it reaches such happiness [as is possible] by sharing a common good....[35]

It is in part Augustine's view of time (of the eternal shadowing the ephemeral) that makes his exegesis possible and makes "medieval" readings of literary texts possible, in turn making medieval texts somewhat different from modern ones. It is a reminder that while Chaucer can seem very modern at times—fragmented, secular—his characters exist in a

[32] Augustine, *City of God*, XIV.vi.
[33] Ibid., XIV.i.
[34] Ibid., XV.ii.
[35] Ibid., XV.iv.

universe shaped by universal history as the medieval world understood it. The pilgrims may be lost, but they are not waiting for Godot; they are waiting for God, even if they do not always acknowledge it at every moment. The Reeve and the Pardoner (in their Bataille-like despair) and the Host (in his Bataille-like joy of poetry and laughter) stand out precisely because they either are unaware of or unwillingly accept that larger structure of which they are a part, the universal narrative into which they literally and all human beings figuratively have been written. In fact, they are in the presence of and participating in the eternal City. One of the most Augustinian moments of the *Canterbury Tales* is when, just before the *Manciple's Tale*, the drunken Cook falls off his horse and several unnamed pilgrims collectively help him into his saddle again:

> And with this speche the Cook wax wroth and wraw,
> And on the Manciple he gan nodde faste
> For lakke of speche, and doun the hors hym caste,
> Where as he lay, til that men hym up took.
> Allas, he nadde holde hym by his ladel!
> And er that he agayn were in his sadel,
> Ther was greet showvyng bothe to and fro
> To lifte hym up, and muchel care and wo,
> So unweeldy was this sory palled goost.
> And to the Manciple thanne spak oure Hoost:
> ….
> "He hath also to do moore than ynough
> To kepen hym and his capul out of the slough;
> And if he falle from his capul eftsoone,
> Thanne shal we alle have ynogh to doone
> In liftyng up his hevy drunken cors.
> Telle on thy tale; of hym make I no fors."[36]

There is comedy here at the Cook's expense, and the Host is somewhat annoyed at the interruption. But at this moment the pilgrims are creating a fellowship otherwise often sorely absent on the road to Canterbury; they are citizens of the City of God. The dead weight of the drunken Cook is "unweeldy" and requires "greet showvyng bothe to and fro /To lifte hym up, and muchel care and wo"—but the pilgrims accept this labor (a pointed echo of the venal Pardoner's earlier rejection of labor). Moreover, the Host emphasizes that the whole group ("we all") has the responsibility to put the Cook back on his horse should he fall again—and though the

[36] *The Manciple's Prologue*, ll.46-68.

Host says this out of irritation (with the pilgrim who is his natural enemy in their economic and social circle), the implication is that the pilgrims *will* help the Cook again, even if it means they will have "ynogh to doone." Their act of kindness, of fraternal love, towards this "sory palled goost" is fleeting and in terms of human history and historical time, insignificant—and yet full of meaning and a symbolic connection to that eternal City which will manifest at the end of time.

Etienne Gilson, at the end of his essay on *City of God* (in which he makes the case for Augustine's opus being one of the earliest and most important universal histories), suggests that Augustine succeeds where later universal histories have failed because modern theories lack a unifying principle, in this case a Catholic theology:

> The desire of the world-wide unity which fills the heart of man will, in all likelihood, never die...Generation after generation has honestly attempted to gather all men within the walls of an earthly city modeled upon the heavenly Jerusalem. They have studied everything except the Christian faith in order to find a common bond, but they have met with failure...It is completely useless to pursue a Christian end except by Christian means. If we really want one world, we must first have one Church, and the only Church that is one is the Catholic Church...Basically, there is nothing wrong in attempting to achieve philosophical unity by philosophical means, nor in trying to establish world unity through philosophical unity. Philosophy really is a unifying force, as are science, art, industry and economic forces...But, just as every metaphysical undertaking is doomed to failure if secondary principles replace those which are primary, so also all efforts to unify mankind are bound to fail if the sole principle of unification is overlooked...Philosophy, science, art and economics all can help in achieving the great work of uniting mankind, but neither individually nor collectively is it in their power to accomplish it...It is conceivable that a number of men, more or less large, be unified under the domination of other men or even of one individual; however, if we are striving toward the unity of all men, we must look beyond mankind for the unifying principle. The only possible source of future unity lies not in multiplicity, but above it. *One World* is impossible without *One God* and *One Church*. In this truth lies the ever timely message conveyed to man by St. Augustine's *City of God*.[37]

Gilson here may be thinking not just about science in the absence of religion or about Marxist economics, but about Hegelian universal history, which has a metaphysical component but which fails to identify a single

[37] Introduction to *City of God*, 34-35.

religion (that is, Catholicism) as the mechanism which will bring about the end of history. Hegel privileges philosophy (as the mediator which will synthesize the opposites of religion and Enlightenment skepticism),[38] and seems to leave room for the possibility of a multiplicity of religions working together dialectically to bring about the union of the temporal world and the divine.[39] Earlier, I focused on Kojève's reading of Hegel partly because it leads to the writings of Bataille and Fukuyama, all secular views of history (their secularism is obviously one of the aspects which marks the boundary between their modernism and the medievalism of Augustine). But Hegel's lectures on religion are interesting in that they elaborate a Christian eschatology much like Augustine's work, and yet the spirit of Hegel's concept of divine history is different.[40] In some ways it is the same difference between Augustine and the other moderns like Kant and Bataille: missing in Hegel is the sense that the individual has a special and meaningful role to play in the grand history (in this case, the process by which the world and the Spirit will be united) unfolding around him.

Gilson may or may not be right about what is necessary to bring about unity in the future and end the history of divisions (that is, to end history). He is right that one element which marks Augustine's universal history is his carefully elaborated theology. But setting aside the question of whether or not it is theology which makes his theory more tenable than modern theories, I think the more interesting point, in terms of navigating the boundary between medieval and modern thought, is that we see Augustine's medieval universal history positioning the individual in history in a different manner, addressing the question of the individual's

[38] Peter C. Hodgson, *Hegel and Christian Theology: A Reading of the Lectures on the Philosophy of Religion* (Oxford, 2005), 200-204.

[39] A dispersal of God throughout the world, as it were. See Hodgson, *Hegel and Christian Theology*, who feels that although Hegel privileges Christianity, there is a tension in Hegel's lectures on theology which make it unclear how Christianity (a historical phenomenon, existing in history) can be the consummate religion (76-77); and who argues that the concept of a multiplicity of religions engaging one another dialectically (in order to help manifest the Spirit in this world) fits better with Hegel's dialectical thinking (see 237-238).

[40] Interestingly, Hegel does not engage Augustine directly in his lectures on religion. Hodgson suggests that while there are similarities in their metaphysics (in his view, perhaps more so than between Hegel and Aquinas), there is no evidence that Hegel was directly familiar with Augustine's theology. See Hodgson, *Hegel and Christian Theology*, 134 and 184, n.10.

place in that history in a more optimistic tone, than many of the universal histories that followed his.

Bataille, looking into the face of Hegel, could only imagine that he saw despair:

> [I]n a portrait of him [Hegel] as an old man, I imagine seeing exhaustion, the horror of being in the depths of things—of being God...[P]erhaps linking the certainty of having attained absolute knowledge with the completion of history—with the passing of existence to the state of empty monotony—he saw himself, in a profound sense, becoming dead; perhaps even his various bouts of sadness took shape in the more profound horror of being God.[41]

In a portrait of Augustine's late life, one imagines seeing only vitality, that reinvigorating relief which comes from not having to be God, yielding the responsibility for the final valuation of one's time in the world, and that of all men, to God.

This Augustinian perception of time, of history coexisting with shadowed eternity, is at the root of the difference between medieval and modern ways of reading and exegesis, but also at the root of the difference between the medieval and the modern individual's perception of their place in the world. If we follow the logic of Augustine to its end, it is not just that Biblical figures or the supermen in history lead lives of symbolic meaning and deep connection to the divine plan, but that any individual life has the potential to be read as a symbol, to be deeply connected.

There must be some comfort in being like a character in a book: leading a complex (often confusing) life, and yet knowing that there is a superstructure bearing one up, a purpose to one's existence, to one's suffering in time, even if one is only a minor character. Post-enlightenment theories of history look for meaning and order to help make sense of the world: Kant insists that there must be such order because it is too disturbing to consider the alternative. But the theories which provide a sense that humanity (if not individual human beings) is making progress toward some certain end, often only provide a purpose for supermen (if even for them); and that leaves the average individual uncertain of their role in the achieving of that end or, that end once realized, of their purpose after it is achieved. The desire both of Kant (looking to the future) and of Bataille (looking back from the end of history) is for a sense of purpose, for a sign of significance, for reassurance that our time in the world is not

[41] Bataille, *Inner Experience*, 110.

wasted. That desire is answered in Augustine's theory of time and of history, where the significance of one's place in time, like that of a character in a book, is always fathomable, if one can read the shadows carefully and thoughtfully.

Bibliography

Alfred the Great. *Alfred the Great: Asser's Life of King Alfred and Other Contemporary Sources*, trans. Simon Keynes and Michael Lapidge (New York, 1983).

Augustine. *City of God*, ed. Vernon J. Bourke with introduction by Etienne Gilson (New York, 1958).

—. *Confessions*, trans. R.S. Pine-Coffin (New York, 1961).

Bataille, Georges. *The College of Sociology 1937-1939*, ed. Denis Hollier (Minneapolis, 1988).

—. *Inner Experience*, trans. Leslie Anne Boldt (Albany, 1988).

Breisach, Ernst. *Historiography: Ancient, Medieval, and Modern* (Chicago, 2007).

Chaucer, Geoffrey. *The Riverside Chaucer*, 3rd edition, ed. Larry D. Benson (Boston, 1987).

Dale, Sharon, Alison Willams Lewin, and Duane J. Osheim, eds. *Chronicling History: Chroniclers and Historians in Medieval and Renaissance Italy* (University Park, PA, 2007).

Fox, Rory. *Time and Eternity in Mid-Thirteenth-Century Thought* (Oxford, 2006).

Fukuyama, Francis. *The End of History and the Last Man* (New York, 2006).

Hodgson, Peter C. *Hegel and Christian Theology: A Reading of the Lectures on the Philosophy of Religion* (Oxford, 2005).

Kant, Immanuel. *Idea for a Universal History from a Cosmopolitan Point of View*, trans. Lewis White Beck (1963). http://www.marxists.org/reference/subject/ethics/kant/universal-history.htm.

Kojève, Alexandre. *Introduction to the Reading of Hegel* (Ithaca, 1969).

Landes, Richard. "The Fear of an Apocalyptic Year 1000: Augustinian Historiography, Medieval and Modern", *Speculum* 75 (2000), 97-145.

Markus, R.A. *Saeculum: History and Society in the Theology of St. Augustine* (Cambridge, 1970).

Pater, Walter. *The Renaissance*, ed. Adam Phillips (Oxford, 1986).

Popper, Karl. *The Open Society and its Enemies Vol. 2: Hegel and Marx* (Princeton, 1966).

Rosemann, Philipp W. "Heidegger's Transcendental History", *Journal of the History of Philosophy* 40 (2002), 501-523.

Smith, D. Vance. "Irregular Histories: Forgetting Ourselves", *New Literary History* 28 (1997), 161-184.

Spiegel, Gabrielle M. *The Past as Text: The Theory and Practice of Medieval Historiography* (Baltimore, 1997).

Thurlow, P.A. "Augustine's *City of God*, Pagan History and the Unity of the Annolied", *Reading Medieval Studies* 6 (1980), 44-67.

CHAPTER TWO

THE MANY FACES OF PETER LOMBARD: CHANGING PERCEPTIONS OF A MASTER IN IMAGES MADE BETWEEN 1150 AND 1215

LAURA CLEAVER

In the thirteenth century a note was added to a chronicle entry for the 1150s, identifying Peter Lombard as one of the leading teachers of his time.[1] His abilities as a scholar contributed to his being made bishop of Paris in 1159, a post he held until his death in 1160.[2] The monument erected to his memory in the thirteenth century celebrated both his writings and his role as bishop, as it declared: "Here lies Peter Lombard, bishop of Paris, who composed the books of the Sentences, gloss on the Psalms and on the Epistles and died on the 13 day before Kalends August [July 20th]".[3] Yet at the time of Peter's death his reputation was not

---

My work on images of teachers has been greatly encouraged and helped by conversations with Patricia Stirnemann. I am grateful to Constant Mews and Clare Monagle for showing me their work in progress, commenting on this paper in draft, and for their insights on the subject of Peter Lombard and heresy. Thanks are also due to Julian Luxford, Michael Clanchy and John Lowden for discussions of this work during its development.

[1] "Roberti Autissiodorensis Chronicon", ed. O. Holder-Egger, *Monumenta Germaniae Historica: Scriptorum* 26 (1882), 223-224 and 237.

[2] M.L. Colish, *Peter Lombard* (Leiden, New York and Köln, 1994), 23.

[3] HIC IACET MAGISTER PETRUS LOMBARDUS PARISIENSIS EPISCOPUS QUI COMPOSUIT LIBRUM SENTENTIARUM GLOSSAS PSALMORUM ET EPISTOLARUM CUIUS OBITUS DIES EST XIII KALENDAS AUGUSTI; Peter Lombard *Sententiae in IV Libris Distinctae: Prolegomena*, ed. I. Brady vol.

secure. Thus when Roger of Wendover, a monk at Saint Albans, subsequently reflected on Peter's career he chose to concentrate on the charges of heresy levelled at Peter's work.[4] He referred to a booklet written at the turn of the thirteenth century by Abbot Joachim of Fiore, in which Joachim declared that Peter was a heretic and insane, and that statements in the *Sentences* were false.[5] Joachim was not alone in his criticism of Peter; from the 1140s Gerhoch of Reichersberg had been suspicious of Peter's work, around 1158 Robert of Cricklade referred to Peter as a heretic, and John of Cornwall and Walter of Saint Victor, writing in the early 1160s and later 1170s respectively, both claimed that Peter's views on Christ's humanity (found in the third book of the *Sentences*) were erroneous.[6] The concern about Peter's work was largely preserved in sources connected with Pope Alexander III and the Papal court, but it had a much wider impact. Roger of Wendover cited a letter of 1170 from the Pope to William, Archbishop of Sens, in which William was ordered to gather his supporters and insist that a doctrine about Christ's humanity, which the Pope identified with Peter, was not to be taught or studied in the schools of Paris.[7] This was followed by a second letter in 1177 (by which time William was archbishop of Reims), in which the Pope ordered William to reiterate to the masters of Reims and Paris that this doctrine was erroneous, though this time he did not mention Peter by name.[8]

As Constant Mews and Clare Monagle have recently argued, these debates probably spread further through gossip and rumour.[9] The heresy attributed by some to Peter was discussed at the Council of Tours in 1163 and probably again at the third Lateran Council of 1179, which were attended by scholars and church authorities from across Europe.[10] Walter

1.1 (Grottaferrata, 1971), 43* (hereafter Brady).

[4] Roger of Wendover, *Roger of Wendover's Flowers of History*, vol. 2, trans. J.A. Giles (London, 1849), 45-48.

[5] Roger of Wendover, 46.

[6] Colish, 427-432; C.J. Mews and C. Monagle, "Peter Lombard, Joachim of Fiore, and the Fourth Lateran Council", *Medieovo* (forthcoming).

[7] Alexander III, "Epistole [...]", *Patrologia Latina* 200, Col. 685.

[8] P. Glorieux, "Le *Contra quatuor labyrinthos Franciae* de Gauthier de Saint Victor", *Archives d'histoire doctrinale et littéraire du Moyen Age* 19 (1952), 202.

[9] Mews and Monagle, "Peter Lombard".

[10] Glorieux, "Le *Contra*", 201; John of Cornwall, "Eulogium ad Alexandrum III Papam Quod Christus sit Aliquis Homo", *Patrologia Latina* 199, Col. 1043; see

of Saint Victor recorded that when the Pope accused Peter of heresy in preparation for the third Lateran council, his hearers left the council, whilst one Bishop Adam sprang to his defence, speaking as a clerk and head of the schools.[11] Thus it seems that some individuals held strong and conflicting opinions about Peter and his work in late twelfth and early thirteenth centuries.

The discussion of the heresy connected with Peter was officially concluded in 1215, when, according to Roger of Wendover, Pope Innocent III condemned Joachim's book and named Peter in connection with correct doctrine.[12] However, in the context of this long period of hostile debate, the production of a large number of lavishly decorated copies of the *Sentences,* both in the area around Paris and further afield, may be read not only as a comment on the work's usefulness and huge popularity as a theological tool, but also as a statement about Peter's worth. Moreover, the decoration of at least ten of these books included images of Peter, which together with depictions of him in copies of his other works make him the most represented near-contemporary author in surviving manuscripts from the period.[13] The images provide further evidence for attitudes towards the master and his work at this time, although they only represent the views of those who wanted decorated copies of his work. Unfortunately, it is very difficult to indentify the dates and places of origin of these books precisely. However, the variation in the images indicates that patrons and artists across France and beyond its borders literally reshaped Peter's identity for their own ends.

One of the most striking representations of Peter Lombard appears in a twelfth-century copy of the *Sentences* now in the Médiathèque at Troyes (MS 900, f.1, fig. 1). The image shows Peter seated at a desk in a typical authorial pose recalling images of the Evangelists. However, the details of his costume suggest that he is to be identified as a significant and wealthy contemporary master. In the image he wears decorated shoes, a large ring and an unusual brown hat that extends over his ears. The impression of wealth is heightened by the gilded background to the initial, which also

---

also "Annales Reicherspergenses a 921-1167", ed. G.H. Pertz, *Monumenta Germaniae Historica: Scriptorum* 17 (1861), 471.

[11] Glorieux, "Le *Contra*", 201.

[12] Roger of Wendover, "Flowers of History", 47.

[13] In addition the opening initials of some volumes have been excised. These might also have contained images of Peter Lombard. See Paris Bibliothèque de l'Arsenal MS 445.

marks this book as a luxury object. The depiction of Peter as a wealthy master must be read as an endorsement of his success, but teaching for financial gain was controversial. In 1159, John of Salisbury, an Englishman who had spent much time in Paris, observed that wealth could be achieved through teaching, but he railed against those who taught solely for this end. Of such teachers John declared: "wisdom's only fruit, for them, is wealth".[14] Nevertheless, in the context of the Troyes manuscript this treatment of Peter suggests that he was to be celebrated as a successful twelfth-century master.

Peter's status and importance as a teacher is further indicated in the Troyes manuscript by the inscription on the open book before him, which reads "come to me all you who are thirsty".[15] This is not a direct Biblical quotation, but, as Patricia Stirnemann has observed, by recalling Christ's words the designer suggests that in Peter's work students will find the wisdom they seek and will be directed to Christ.[16] Peter is thus presented as a contemporary authority on scripture, a role justified principally by his status as a teacher. Such a reputation is also suggested in a poem of the 1140s, where "the Lombard" heads a list of teachers as a "famous theologian".[17] Similarly, Peter of Celle, Abbot of Saint Remi at Reims and later bishop of Chartres, wrote to a monk at Saint Bertin c.1174-1181, "If new works please you see the writings of master Hugh, see those of Saint Bernard, of master Gilbert, and master Peter, in which neither roses nor lilies are lacking".[18] It seems probable that he was referring to Hugh of

[14] "Nihil enim sordidum putant, nihil stultum, nisi paupertatis angustias, et solas opes ducunt esse fructum sapientiae", John of Salisbury, *Metalogicon,* ed. J.B. Hall, (Turnholt, 1991), 19-20; translation from John of Salisbury, *The Metalogicon of John of Salisbury: A Twelfth-Century Defense of the Verbal and Logical Arts of the Trivium*, trans. D.D. McGarry (Berkley and Los Angeles, 1962), 20.

[15] "Omnes sitientes venite ad me."

[16] P. Stirnemann, "Histoire tripartite: un inventaire des livres de Pierre Lombard, un exemplaire de ses *Sentences* et le destinataire du Psautier de Copenhague", in *Du copiste au collectionneur: mélanges d'histoire des textes et des bibliothèques en l'honneur d'André Vernet*, eds J-F. Geneat and D. Nebbiai-Dalla Guarda (Turnhout, 1998), 311; see Matthew 11:28 and Isaiah 55:1.

[17] "Celebrem theologum vidimus Lumbardum,
cum Yvone Helyam Petrum et Bernardum,
quorum opobalsamum spirat os et nardum,
et professi plurimi sunt Abaielardum", R.B.C. Huygens, "Mitteilungen aus Handschriften", *Studi Medievali* third series, 3 (1962), 771; Colish, 20.

[18] Peter of Celle, *The Letters of Peter of Celle*, ed. J. Haseldine (Oxford, 2001),

Saint Victor, Bernard of Clairvaux, Gilbert of Poitiers and Peter Lombard, though in praising them Peter of Celle ignored the fact that three of these men had challenged each other's views on theological matters.[19]

The term "new" is significant in Peter of Celle's letter, as it distinguishes Peter and his contemporaries from earlier authors. The Troyes manuscript bears the date 1158, and this, together with the fact that Peter is not dressed as a bishop, a role he achieved in 1159, contributed to Stirnemann's suggestion that 1158 might be the date of production for this manuscript in Paris, thus making it contemporary with Peter himself.[20] However, scholars working on the text have identified the date as that of the completion the *Sentences*, rather than of this copy.[21] The precise dating of manuscripts from the second half of the twelfth century remains controversial, and the style of the images in the manuscript could be in keeping with a slightly later date.[22] Moreover, this is not the only volume not to present Peter as a bishop. An image executed in a comparable style appears in another copy of the *Sentences* probably made in Paris, now in Bamberg (Staatsbibliothek Msc. Patr. 120).[23] The initials in both manuscripts are surrounded by areas of a single colour decorated with white dots and lines. Moreover, the curling ends of the letter sprouting heads in the Troyes manuscript reappear in the initial to the first book in the Bamberg version.[24] In the latter copy Peter is shown writing at a desk, wearing a simple black tunic with a yellow sleeve visible beneath it. He is probably supposed to be tonsured, though the top of his head is cut off by the shape of the letter. Yet despite this simple and ambiguous dress, the open page before him bears the letters M P L E for *magister Petrus Lombardus Episcopus*. The manuscript must thus have been made after Peter became a bishop in 1159. Whether the Troyes volume was executed in Peter's lifetime, or shortly after, it may have stood at the beginning of a

555-557, n.5.

[19] L. Cleaver, *Art and Education in Northern Europe c.1080-c.1220* (PhD thesis, University of London, 2008), 130-173.

[20] Stirnemann, "Histoire tripartite", 301-318.

[21] Brady, 125*, 130*-131*.

[22] Stirnemann, "Histoire tripartite", 308.

[23] G. Suckale-Redlefsen, *Katalog der illuminierten Handschriften der Staatsbibliothek Bamberg: Die Handschriften des 12. Jahrhunderts*, vol. 2 (Wiesbaden, 1995), 82-84.

[24] Suckale-Redlefsen, Tafel V.

tradition of depicting Peter as an author and teacher without referring to his status as a bishop.

The Bamberg and Troyes manuscripts were certainly not executed by the same artist, but the similarities between them extend to the iconography of other initials. Both books have images of Synagogia and Ecclesia in the initial to Book 1, the Nativity at Book 3 and the Good Samaritan at Book 4. The Troyes version also depicts God with Adam and Eve at the prologue to Book 2, though the Bamberg version does not include figures in this initial.[25] Part of this iconographic sequence also appears in another twelfth-century copy of the *Sentences,* now in the library of St. John's College, Oxford (MS 49), which may thus be associated with the Troyes and Bamberg volumes.[26] Indeed, Stirnemann identifies this copy as having been decorated by the same artist as that of the Troyes volume.[27] The Oxford manuscript includes an image of the creation of Eve at the start of Book 2 and the Nativity at Book 3. The initial to Book 4 does not include figures, and that at Book 1 has been cut out. Whilst there are strong stylistic similarities between the two volumes, that in Oxford has been executed rather more simply, for example using a smaller initial for the Nativity scene, in which the curtain, lamp and foliate scrolls of the Troyes version do not appear. However, notably the Oxford volume does not include an image of Peter Lombard in the opening initial, which is simply executed and lacks the gilding of the other initials. Despite the difference in the figures of the author in the Troyes and Bamberg volumes, therefore, these three manuscripts suggest that a set of images was repeatedly used for copies of the *Sentences* made in Paris in the second half of the twelfth century, and the comparison makes clear that the inclusion of the author figure was a deliberate choice. Moreover, the Oxford volume provides a compelling indication that the artist was active before 1170, as a later note records that it was the gift of Hilary, Bishop of Chichester, who died in 1169.[28] Thus it seems likely that the Troyes volume was produced either in the final years of Peter's life, or shortly after his death.

---

[25] See http://www.enluminures.culture.fr for images of the Troyes manuscript [accessed November 2009].

[26] R. Hanna, *A Descriptive Catalogue of the Western Medieval Manuscripts of St. John's College Oxford* (Oxford, 2002), 68-69.

[27] Stirnemann, "Histoire tripartite", 309.

[28] Hanna, *A Descriptive Catalogue*, 69.

Given the similarity in the decorative schemes in the Troyes and Bamberg manuscripts, the different presentations of the figure of the author in the two volumes are striking. Moreover, in the case of the Troyes manuscript the costume worn by Peter is unusual. The master wears a long white tunic, caught up to show off his leg, and a black cowl with a hood. It is conceivable that the Bamberg volume's black tunic, decorated with white dots in groups of three, with a yellow sleeve visible beneath is a variation on this, but if so it is much less specific, instead resembling generic dress found in other author portraits.[29] Yet despite the detail in the Troyes image it is not an accurate portrait of Peter as he was in life. Black cowls and white tunics were worn by Augustinian canons, or Canons Regular, represented in Paris by the community just outside the city walls at Saint Victor.[30] Such a reading of this image might support Stirnemann's suggestion that this volume was made in Paris for the Victorines and subsequently given to the Cistercians, making its way to the library at Clairvaux, where it is recorded in the fifteenth century.[31] If so, however, the dress would reflect that of the manuscript's reader, but not of Peter himself, who was based at the cathedral of Notre-Dame, which was not staffed by Augustinians. In a letter of c.1134-36, Bernard recommended Peter to Gilduin, Abbot of Saint Victor in Paris, and noted that he had helped to fund Peter's studies in Reims.[32] However, whilst Peter may have lodged briefly at Saint Victor, he did not join that community. Thus the designer of this image has shaped an alternative identity for Peter, presenting him firstly as a contemporary master, but also potentially linking him to the Augustinians and the religious life.

A rather different image of Peter Lombard, probably made for the library of the Benedictine house at Montier-Ramey near Troyes, may be found in another copy of his *Sentences* in Troyes (Médiathèque MS 304 f.1, fig. 2). This volume also has an image of Synagogia and Ecclesia at the start of Book 1, though the other initials do not contain figurative imagery.[33] The image of Peter is in poor condition, perhaps as a result of

---

[29] See, for example, Admont Monastery Library MS 36 f.1, and Troyes Mediathètheque MS 48 f.1 both described below.

[30] See also Paris Bibliothèque Nationale de France MS Latin 14515 f.106v and Oxford, Bodleian Library MS Laud Misc 409 f.3v.

[31] Stirnemann, "Histoire tripartite", 313.

[32] Bernard of Clairvaux, *Opera*, vol. 8 (Rome, 1977), 391; Colish, *Peter Lombard*, 16-17.

[33] See http://www.enluminures.culture.fr for additional images [accessed November

the book's frequent use. Peter appears in the opening initial to the text and is presented on a gilded background. He holds up an open book, but is not shown writing in it. Here he is again without bishop's regalia, wearing a red hooded cloak over a long green tunic. He has no hat and appears to be tonsured. Despite the same combination of tunic and cowl, therefore, in contrast to the image in the volume dated 1158, he is presented as a modest figure, rather than as a fashionable master. This does not seem to be the result of a lack of ambition on the part of the artist, given the use of a gilded background and the other figurative initial. Instead this variation in the imagery may have been thought a more appropriate representation of a theologian by or for an audience of Benedictine monks. The instructions of a patron on this point must remain speculation, but by making Peter a recognisable figure the images may have helped viewers to identify themselves with the master and to consider themselves as his students, albeit at a remove. Such a reading seems to be supported by the open book held by Peter in the Montier-Ramey manuscript, which mirrors the form of the volume open before the reader, just as the volume in the copy dated 1158 invites readers to go to Peter for instruction.

The possibility that author portraits could serve to help the reader identify himself as a student of a master is also suggested by an image in a manuscript made c.1200, probably for Saint Albans, and now in Oxford (Bodleian Library MS Laud Misc 409 f.3v). The image occupies a full page and prefaces Hugh of Saint Victor's treatise *On Noah's Ark*, which begins, "When one day I was sitting in the assembly of brothers and responding to the questions they asked me many matters were discussed" and ends with a suggestion that the students are tired.[34] As Michael Evans explored,[35] the descriptions are a literary device, as the text is here being written and read with readers free to pause when tired, but Hugh's words evoke the classroom as the context in which the work is to be read. In the same way, the image shows Hugh dressed in the black and white of an Augustinian canon and seated on a large chair opposite three small pupils, again associating the work with a classroom debate. Moreover, both Hugh and the foremost student hold open books that are displayed facing the

2009].

[34] Hugh of Saint Victor, *Hugonis de Sancto Victore: De Archa Noe, Libellus de Formatione Arche*, ed. P. Sicard (Turnhout, 2001), 3, 117; M. Evans, "Fictive Painting in Twelfth-century Paris", in *Sight and Insight: Essays on Art and Culture in Honour of E.H. Gombrich at 85*, ed. J. Onians (London, 1994), 82.

[35] Ibid.

viewer and thus resemble the open book before the reader. In this case it seems that the viewer is invited to include himself amongst Hugh's pupils, and the same may be true of the images of Peter Lombard. In the Montier-Ramey manuscript, Peter holds the book with his hand covered, indicating its importance and perhaps also emphasising the connection between his work and holy scripture. However, Peter's text begins, "Wishing, with the poor widow, to give something to the Lord's treasury out of our penury and poverty, we have dared to scale the difficult heights and undertake a work beyond our strength".[36] This opening seems thus to conflict with the gilded presentations of Peter and particularly with the finery of the figure at the start of the volume dated 1158. Whilst the Montier-Ramey image may in part be a response to these lines, therefore, both artists opted to present Peter in a more vaunted guise than he adopted for himself, albeit in different clothing.

The relative abundance of particular detail in the two images now in Troyes is thrown into relief by at least three other images of Peter in copies of his *Commentary on the Psalms* from late twelfth and early thirteenth centuries. In a volume produced in the 1160s for the Bishop of Salzburg and now in the Benedictine monastery library at Admont (MS 36 f.1) the author is shown seated on an architectural throne, holding up a rectangular object, probably a book.[37] He has a tonsure and a loose-fitting tunic decorated with dots in groups of three, with black shoes, and is thus identified as a cleric and author, though like the figure in the Bamberg volume he is otherwise undistinguished by his dress. Similarly, the opening initial in a manuscript probably made in England in the 1160s or 1170s now in the British Library (Additional MS 54229 f.3), shows two figures, a master and a student, both of whom are dressed simply in tunics and cloaks.[38] Indeed, given that this text is a commentary on the Psalter it is unclear whether the master should be interpreted as Peter or David. If the latter, Peter might be represented by the figure of the student, absorbing and transmitting David's work. The image resembles scenes of Solomon teaching, which are often found in twelfth-century Bibles, and although the master is not crowned, he is not tonsured, making it uncertain

[36] Peter Lombard, *The Sentences: Book 1 The Mystery of the Trinity*, trans. G. Silano (Toronto, 2007), 3.
[37] C. de Hamel, *Glossed Books of the Bible and the Origins of the Paris Booktrade* (Woodbridge, 1984), 9; Cleaver, *Art and Education*, fig. 4.24.
[38] http://www.bl.uk/onlinegallery/onlineex/illmanus/other/011add000045229u000 03000.html [accessed August 2009].

whether he is a medieval or an ancient author.[39] A third example, probably made in France at the turn of the thirteenth century and now also in Troyes (Médiathèque MS 48 f.1), is a little more precise, as here Peter is tonsured and identified by the book before him which bears his name.[40] However, here he wears a green cloak, without a hood, over a red tunic decorated with groups of three dots, again providing minimal information about his status. Nevertheless, the use of costly materials in this initial, which has a gilded background, again indicates the importance of Peter and of this work, which was never accused of heresy.

In contrast, other images of Peter stress his status as a bishop apparently in order to emphasise his orthodoxy. Thus in two copies of the *Sentences* he is presented in a bishop's robes and with a halo, despite the fact that he was never canonised. The better known of these examples came from the library of the Benedictine monastery at Saint Amand in northern France and is now held at Valenciennes (Bibliothèque Municipale MS 186, f.2v).[41] This manuscript, probably made in the 1160s, when the group associated with Paris was also active, shows Peter standing, carrying a crosier and with one hand outstretched as if to indicate that he is speaking. This gesture associates the work that follows with Peter's preaching as bishop, stressing the ecclesiastical status of the man who produced these words. Similarly, the halo indicates that Peter is dead, and likens him to the saints in heaven, further suggesting that his words are likely to be approved by God, if not necessarily by the leaders of the church on Earth. The book is large, measuring 43.7 by 29.5 cm, and the artist is identified in the lines beneath the figure as Sawalo, a monk of Saint Amand. Such a format was more commonly used for volumes of works by the Church Fathers, thus the scale of this manuscript together with the size of this image, which is the only figurative scene in the volume, stresses the author's worth in the eyes of the artist who linked his own name so prominently with Peter.

A similar image of Peter as a standing bishop was included in another large manuscript of the *Sentences*, measuring 43.7 by 31.5 cm, now in

[39] C. Mattke, "Verges et disciplines dans l'iconographie de l'enseignement", *Médiévales (Du bon usage de la souffrance)* 27 (1994), 107-120; Cleaver, *Art and Education*, 30-70.

[40] See http://www.enluminures.culture.fr [accessed August 2009].

[41] A. Legner, ed., *Ornamenta Ecclesiae: Kunst und Künstler der Romanik* vol. 1 (Cologne, 1985), 254; *Catalogue général des manuscrits des bibliothèques publiques de France: Départements: Valenciennes* vol. 25 (Paris, 1894), 270.

Reims (Bibliothèque Municipale MS 460 f.157v, fig. 3).[42] The volume bears the ownership mark of the Abbey of Saint Denis in Reims and may be dated to the end of the twelfth century. Peter is again shown wearing bishop's robes, carrying a crosier and haloed, though here he has grey hair, perhaps to indicate his maturity. However, in this volume the image is not placed at the start of the text, the usual position for author portraits, but at the prologue to Book 3. The location of the rather conventional figure of a bishop at this point makes it clear that this is an image of Peter Lombard, as this part of the text, like the prologue to Book 1, was written in the first person. Moreover, this was also the start of the section accused by some of heresy, as it deals with the nature of the Trinity and Christ's humanity. It is thus conceivable that the decision was made to place the figure of Peter at this point to remind the reader of his ecclesiastical authority at the place where his orthodoxy might be challenged.

Peter is also identified as a bishop at the start of the first book of the *Sentences* in a late twelfth-century copy in Paris (Bibliothèque Nationale de France MS Latin 14513, f.1).[43] The initial is painted and the letter itself is gilded, though the volume as a whole is relatively modest and measures 33 by 24 cm. In this case Peter does not have a halo, but he holds a slender white crosier and points to the start of the text, again identifying the volume as having been written by a man in a high ecclesiastical office. This manuscript was probably made in England, indicating the rapid spread of interest in Peter and his work also evidenced by Roger of Wendover's history.[44]

In addition to the images of Peter as a haloed bishop, his orthodoxy was also established through images that showed him as a divinely inspired scribe. In a manuscript now in Vienna (Österreichische Nationalbibliothek MS Lat. 1402, f.8r), in a copy of the *Sentences* made in France at the end of the twelfth century, Peter appears in the initial to the start of Book 1.[45] He is again identified as a bishop, with a pointed mitre, and his religious authority is further illustrated by the presence of the Holy

[42] *Catalogue général des manuscrits des bibliothèques publiques de France: Départements: Reims*, vol. 38, ed. H. Loriquet (Paris, 1904), 629-630.

[43] G. Ouy, *Les Manuscrits de l'Abbaye de Saint-Victor*, vol. 2 (Turnhout, 1999), 77-78.

[44] F. Avril and P. Stirnemann, *Manuscrits enluminés d'origine insulaire VIIe-XXe siècle* (Paris, 1981), 38-39.

[45] J. Hermann, *Die romanischen Handschriften des Abendlandes mit Ausnahme der deutschen Handschriften* (Leipzig, 1927), 30.

Spirit in the form of a dove, here shown speaking into his ear.[46] This detail is repeated in a late twelfth-century copy of Peter's *Commentary on the Epistles* from Saint Victor, which remains in Paris (Bibliothèque Nationale de France MS Latin 14266) in which Peter and Paul sit together, but with Peter placed on the left, the place usually given to a master, and holding a scroll that bears his name.[47] In this case Peter is presented not only as the inheritor of Paul's work, but also as being inspired in his own work by God. These images thus mark a subtle shift of attitude from that of the manuscript dated 1158, where Peter was shown as a modern master providing his own commentary on the scriptures, and instead present him as a channel through which God's ideas may be known.

The initials in the Paris and Vienna manuscripts both identify Peter as a teacher in the presence of students. In the Vienna manuscript Peter is joined by a small pupil, seated at his feet, who receives a scroll from him. In the Paris initial the role of Peter and Paul as transmitters of knowledge is suggested by the presence of a crowd of smaller figures in a circle forming part of the stem of the letter. These figures look towards the authors and hold up their hands, which are covered, thus suggesting great reverence. The distance between these students and their masters indicates that the teachers are now dead and known only through their writings. However, the inclusion of the students may also suggest that the Pope's edicts in restricting teaching connected with Peter had a limited impact. The imagery certainly suggests that the designer of the initial considered the text a useful tool for students, perhaps including those who used these volumes. The high cost of producing decorated volumes makes it unlikely that they were intended for use by students in a lay classroom, but they might have been pored over by those using them in monastic libraries.

The importance of links between individual masters and students, as well as the growing identification of different teaching with particular schools is recorded in some of the writings about Peter's heresy. Thus in his treatise against him, John of Cornwall identified himself as one of

---

[46] J.P. Turcheck, "A Neglected Manuscript of Peter Lombard's *Liber Sententiarum* and Parisian Illumination of the Late Twelfth Century", *The Journal of the Walters Art Gallery* 44 (1986), 57.

[47] L.M. Ayres, "A Miniature from Jumièges and Trends in Manuscript Illumination around 1200", *Intuition und Kunstwissenschaft: Festschrift für Hanns Swarzenski zum 70. Geburtstag am 30. August 1973*, eds P. Bloch et al. (Berlin, 1973), 122-123; Turcheck, "A Neglected Manuscript", 58; R. Branner, "Manuscript Painting in Paris around 1200" in *The Year 1200: A Symposium* (New York, 1975), 174.

Peter's students, claiming knowledge through this connection, although he separated himself from Peter by denouncing his work.[48] However, the writer and archbishop William of Tyre testified to Peter's talent as a teacher, noting that he had studied with him for six years.[49] Moreover, as Martha Colish has noted, the continued debates on the subject of the heresy associated with Peter may also have been connected to the teaching of Peter of Poitiers, a follower of Peter Lombard's, who taught theology in Paris from 1163 and was a prominent figure.[50] Accusations of heresy in an individual's work could have consequences for the students of that master, particularly if they sought to defend the controversial statements. Such links take visual form in an image in another manuscript from Saint Amand (Valenciennes, Bibliothèque Municipale MS 197, f.4v), which contains a copy of Gilbert of Poitiers' controversial commentary on Boethius' *De Trinitate*, a work which was accused of heresy in the 1140s. In a full-page image Gilbert is shown passing his teaching in the form of a scroll to three named students, whilst a fourth recipient of his wisdom appears writing at the top of the following page. Thus in this case the designer of the image seems to have wanted to stress the value not only of Gilbert's work on this controversial subject, but also of those who studied it.[51] However, unlike Gilbert's commentary, whilst Peter's work was much debated, the large number of surviving copies from the twelfth century indicates that it continued to be very widely used by students who held Peter's thoughts to be in line with scripture.

The idea that Peter's work was worth incorporating into the canon of approved theological sources is further suggested by the repetition of particular iconographic forms for Peter and other authors. Thus the artist of the Valenciennes manuscript used the same formula that he used to depict Peter for an image of Saint Hilaire, the fourth-century bishop and writer, in a manuscript now in Paris (Bibliothèque Nationale de France MS

---

[48] John of Cornwall, Col. 1053.

[49] "In theologia autem virum in ea scientia singularem, cuius opera que exstant prudentum chorus cum veneratione amplectitur et colit cum reverentia, virum sana doctrina per omnia commendabilem, magistrum videlicet Petrum Lombardum, qui postea fuit Parisiensis episcopus, annis sex continuis diligenter audivimus", R.B.C. Huygens, "Guillaume de Tyr étudiant: Un chapitre (XIX, 12) de son 'Histoire' retrouvé", *Latomus* 21 (1962), 823.

[50] Colish, *Peter Lombard*, 435.

[51] L. Cleaver, "Heretic or Hero? Posthumous Representations of Gilbert of Poitiers in Words and Images before 1200", *Word and Image* 26 (2010), 285-296.

Latin 1699, f.3).[52] In the inscription beneath this image the man responsible for the book's decoration is again identified as Sawalo, the monk of Saint Amand. It is unclear which image was made first, though the saint is justified in having a halo, and his raised left hand holds an initial, rather than simply displaying an open palm, which may indicate that this image was the earlier and was adapted for that of Peter. If so, the artist only made minimal changes to the iconography of Hilaire, although the inclusion of the halo and raised hand for Peter must nonetheless be read as a deliberate choice. Most of the changes seem to have been influenced by the different layout of the page, which provides a more visually impressive setting for Peter. The master is separated from the large initial and is given greater visual prominence, taking up more than a quarter of the folio. He is placed within a decorated border, but his halo protrudes over the framing device. In contrast, Hilaire is part of an initial and placed in the margin of the text. The saint stands on a dragon, which this time is not part of the initial, further suggesting his righteousness and power. This creature was omitted from the image of Peter, but the adaptation of the iconography thought fit for Hilaire, together with the lavish treatment of the image of the former, suggests a great respect for Peter and his work. Sawalo's use of the same formula for the two men thus indicates that he considered Peter worth including in a canon of major theologians, a remarkable statement about a near-contemporary figure.

Another artist also reused a pattern for an image of Peter and that of an earlier authority. In an image probably made in the 1160s from the monastery of La Charité, near Nevers, now in the British Library (Yates Thompson MS 17, f.42v), Peter is presented as a bishop in mitre and robes, but also as an author as he is seated at a desk and shown writing on a piece of parchment (fig. 4).[53] The image is part of the initial letter to the second book of the *Sentences,* which also begins with a sentence in which the author refers to his work. The open book identifies the figure, since it reads *Petrus Lumbardie eps*. Peter is shown writing on a second sheet, but rather than penning his text or copying the open book before him he has inscribed the letters A to E. These letters recall images of grammar lessons, and the artist may have wished to suggest the role of the bishop's

[52] N. Garborini, *Der Miniator Sawalo und seine Stellung innerhalb der Buchmalerei des Klosters Saint-Amand* (Cologne, 1978), 78.

[53] http://www.bl.uk/catalogues/illuminatedmanuscripts/record.asp?MSID=8119andCollID=58andNStart=17 [accessed August 2009]

work in teaching.[54] Peter's status as an authority is furthered here not only by his bishop's robes, but also by the artist's decision to use the same figure for a portrait of Isidore of Seville (fig. 5). As Sir Sydney Cockerell noted, an almost identical figure of a bishop appears in another manuscript from La Charité, now also in the British Library (Additional MS 15603, f.2).[55] Isidore is placed in an oval that seems to have been adapted from an initial, suggesting that the image of Peter or another bishop may have been the model. The earlier bishop is shown in exactly the same pose as Peter, but the open book bears his name. He is also shown writing letters, though he has only got as far as D. Again, therefore, Peter is presented as the latest in a canon of theological writers, but this time there is also a hint of the importance of his work for teaching.

The idea of a canon of writers, and particularly figures from antiquity, is important in many of these images. Such authors are also evoked by another image of Peter as bishop in a manuscript made c.1180 in the area of Sens, now in the Walters Art Museum in Baltimore (MS W.809, f.1).[56] In the opening initial to this copy of the *Sentences*, Peter is shown seated at a desk wearing a round mitre like that in the Valenciennes volume, but he also wears a piece of cloth wrapped around his body resembling a toga, a garment often found in images of the Evangelists. Peter is shown writing in a book, holding up one hand with two fingers outstretched in a gesture suggesting speech and perhaps also blessing. Thus the image seems again to suggest a dual identity for Peter as a teacher and a bishop. A similar image can be found in a copy of the *Commentary on the Psalms* probably made in the same region and at roughly the same time as the Baltimore volume, now in Bamberg (Staatsbibliothek Msc. Bibl. 60, f.175).[57] Here Peter is shown seated before a blank book, wearing a triangular mitre, but again dressed in a toga-like garment. As in the Baltimore image, the hand

---

[54] Cleaver, *Art and Education*, 30-70.

[55] H. Yates Thompson et al., *A Descriptive Catalogue of the Second Series of Fifty Manuscripts in the Collection of Henry Yates Thompson* (Cambridge, 1902), 130-133.

[56] For the place of origin for these manuscripts see: de Hamel, *Glossed Books*, 38-54; P. Stirnemann, "Review of C. De Hamel *Glossed Books of the Bible and the Origins of the Paris Booktrade*", *Bulletin Monumental* 143 (1985), 362-367; Turcheck, "A Neglected Manuscript", 51, 60-61; P. Stirnemann, "En quête de Sens", in *Quand la peinture était dans les livre: Mélanges en l'honneur de François Avril*, eds E. König, C. Zöhl, M. Hofmann (Turnhout, 2007), 303-311.

[57] Suckale-Redlefsen, *Katalog*, 89-90.

that does not hold writing equipment is raised as if to suggest speech or blessing, but here the decision to place Peter facing the text of the Psalms yet looking towards his commentary results in a twisted pose, rendered more peculiar by the artist's decision to show the face in profile looking away from the book before him. This arrangement of the body was also used for a figure of Paul in another copy of the *Commentary on the Psalms* (Admont Monastery Library MS 53, f.1), where the author was again wrapped in a toga-like cloth.[58] However, the images differ in that Peter was given a mitre in recognition of his Episcopal status. Thus the images seem to indicate another convention for author portraits and for the decoration for volumes of Peter's works in particular, in which he was likened to authors from the ancient world. The combination of the roles of bishop and canonical author again stresses his orthodoxy in line with both scripture and earlier authors.

The creation of these books in the area of Sens in the 1180s places them in a region where the issue of heresy should have been well known. As noted above, in the 1170s the Pope had written to the archbishop of the diocese to order him to stop the teaching of the heresy associated with Peter in the area. The combination of the toga-like garment and bishop's robes in this context is suggestive, as defenders of Peter's work sought to demonstrate his orthodoxy by stressing both his church office and his conformity to earlier writers. Indeed, Peter himself had been aware of the need to work in the tradition of those who had gone before; in the Prologue to the *Sentences* he declared:

> Here you will find the precedents and teaching of our ancestors. Here, by the sincere profession of the Lord's faith, we have denounced the falsehood of a poisonous doctrine. Embracing an approach to showing the truth without incurring the danger of professing impiety, we have pursued a moderate middle course between the two. And if in some places our voice has rung out a little loudly, it has not transgressed the bounds set by our forefathers.[59]

Given the trials of other masters in recent decades, and his own critics, Peter's desire to stress his orthodoxy is unsurprising. However, the decision on the part of those responsible for the decoration of these books

[58] Turcheck, "A Neglected Manuscript", 55.
[59] Peter Lombard, *The Sentences*, 4.

to take up the theme may suggest an attempt to defend the work in the face of particular local challenges.

Thus between the completion of the *Sentences* in 1158 and the Pope's endorsement of Peter in 1215, the master was represented in a wide range of guises in luxury copies of his works. In studying the surviving images of Peter several trends may be discerned. From a very early date Peter was presented by at least one artist as a fashionable contemporary master as part of a series of images in a very costly manuscript. This may have stood at the beginning of a tradition of images of Peter as a master, rather than as a bishop, celebrating him as a contemporary, or near-contemporary, teacher. Other artists simply presented Peter as a generic figure, drawing on a tradition of author portraits to make a comment about the importance of this work and its author. Such images, however, were more common in copies of Peter's other works. In copies of the *Sentences* it became usual to represent Peter as a bishop, and many artists went further, emphasising his church-approved status and linking this to the text. In a similar vein, some artists used imagery also applied to famous theological writers for their images of Peter. In particular, one group of images sought to create a similar association through the combination of bishop's dress and toga-like robes. Books do not need images of their authors, thus the inclusion of these images suggests that Peter was widely celebrated as a theologian and author. However, some images may also have been prompted by a desire to defend the reputation of the master during a period when it was repeatedly challenged. In this process some images seem to invite readers to identify themselves as students of the master, and others to defend those students. In 1215 those who had continued to commission such lavish copies of this book must have felt themselves justified by the outcome of the Lateran Council. Familiarity with the text helped to reveal that it was not in fact heretical, and instead it became an important tool for the next generation of students.

Illustrations

Figure 1, Peter Lombard, Troyes Médiathèque MS 900, f.1, detail, © IRHT-Troyes Médiathèque.

Figure 2, Peter Lombard, Troyes, Médiathèque MS 304, f.1, detail, © IRHT-Troyes Médiathèque.

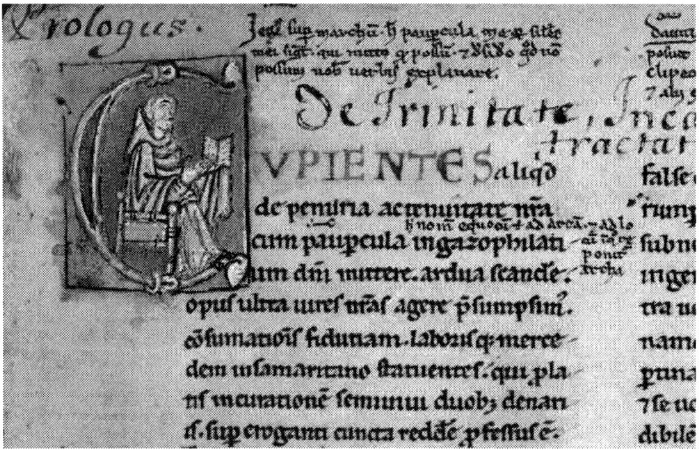

Figure 3, Peter Lombard, Reims Bibliothèque Municipale, MS 460 f.157v, detail, © Reims, Bibliothèque Municipale.

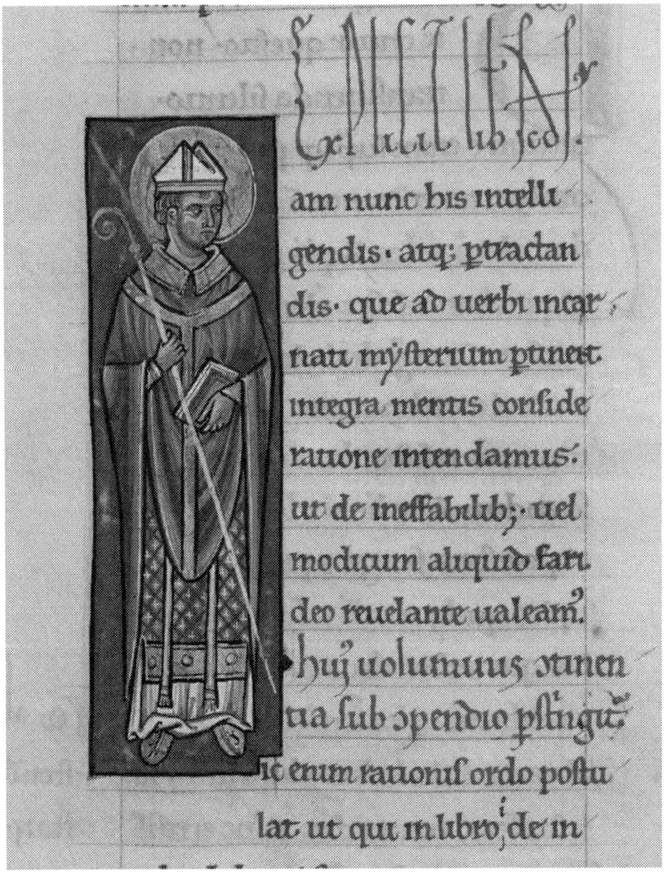

Figure 4, Peter Lombard, British Library, Yates Thompson MS 17 f.42v, detail, © British Library.

Figure 5, Peter Lombard, British Library, Additional MS 15603 f.2, detail, © British Library.

Bibliography

Alexander III, "Epistole [...]", *Patrologia Latina* 200, Col. 685

Anonymous, "Annales Reicherspergenses a 921-1167", *Monumenta Germaniae Historica: Scriptorum* ed. G.H. Pertz 17 (1861), 443-476.

"Roberti Autissiodorensis Chronicon", ed. O. Holder-Egger, *Monumenta Germaniae Historica: Scriptorum* 26 (1882), 219-276.

Avril, F., and P. Stirnemann, *Manuscrits enluminés d'origine insulaire VIIe-XXe siècle* (Paris, 1981).

Ayres, L.M. "A Miniature from Jumièges and Trends in Manuscript Illumination around 1200", *Intuition und Kunstwissenschaft: Festschrift für Hanns Swarzenski zum 70. Geburtstag am 30. August 1973*, eds P. Bloch et al. (Berlin, 1973) 115-139.

Branner, R. "Manuscript Painting in Paris around 1200", in *The Year 1200: A Symposium* (New York, 1975), 175-185.

Peter of Celle, *The Letters of Peter of Celle*, ed. J. Haseldine, (Oxford, 2001).

Bernard of Clairvaux, *Opera* vol. 8 (Rome, 1977).

Cleaver, L. *Art and Education in Northern Europe c.1080-c.1220* (PhD thesis, University of London, 2008).
—. "Heretic or Hero? Posthumous Representations of Gilbert of Poitiers in Words and Images before 1200", *Word and Image* 26 (2010), 285-296.
Colish, M.L. *Peter Lombard* (Leiden, New York and Köln, 1994).
John of Cornwall, "Eulogium ad Alexandrum III Papam Quod Christus sit Aliquis Homo", *Patrologia Latina*, 199, Cols. 1041-1086.
De Hamel, C. *Glossed Books of the Bible and the Origins of the Paris Booktrade* (Woodbridge, 1984).
Evans, M. "Fictive Painting in Twelfth-century Paris", in *Sight and Insight: Essays on Art and Culture in Honour of E.H. Gombrich at 85*, ed. J. Onians, (London, 1994), 73-88.
Garborini, N. *Der Miniator Sawalo und seine Stellung innerhalb der Buchmalerei des Klosters Saint-Amand* (Cologne, 1978).
Glorieux, P. "Le Contra quatuor labyrinthos Franciae de Gauthier de Saint Victor", *Archives d'histoire doctrinale et littéraire du Moyen Age* 19 (1952), 187-335.
Hanna, R. *A Descriptive Catalogue of the Western Medieval Manuscripts of St. John's College Oxford* (Oxford, 2002).
Hermann, J. *Die romanischen Handschriften des Abendlandes mit Ausnahme der deutschen Handschriften* (Leipzig, 1927).
Huygens, R.B.C. "Mitteilungen aus Handschriften", *Studi Medievali* third series, 3 (1962), 764-772.
—. "Guillaume de Tyr étudiant: Un chapitre (XIX, 12) de son 'Histoire' retrouvé", *Latomus* 21 (1962), 811-829.
Legner, A., ed. *Ornamenta Ecclesiae: Kunst und Künstler der Romanik* vol. 1 (Cologne, 1985).
Lombard, Peter. *The Sentences: Book 1 The Mystery of the Trinity*, trans. G. Silano, (Toronto, 2007).
—. *Sententiae in IV Libris Distinctae: Prolegomena*, ed. I. Brady, vol. 1 (Grottaferrata, 1971).
Loriquet, H., ed. *Catalogue général des manuscrits des bibliothèques publiques de France: Départements: Reims* vol. 38 (Paris, 1904).
Mattke, C. "Verges et disciplines dans l'iconographie de l'enseignement", *Médiévales (Du bon usage de la souffrance)*, 27 (1994), 107-120.
Mews, C.J., and C. Monagle, "Peter Lombard, Joachim of Fiore, and the Fourth Lateran Council", *Medieovo* (forthcoming).
Molinier, A., ed. *Catalogue général des manuscrits des bibliothèques publiques de France: Départements: Valenciennes* 25 (Paris, 1894).
Ouy, G. *Les manuscrits de l'Abbaye de Saint-Victor*, vol. 2 (Turnhout, 1999).

Hugh of Saint Victor, *Hugonis de Sancto Victore: De Archa Noe, Libellus de Formatione Arche*, ed. P. Sicard (Turnhout, 2001).
John of Salisbury, *Metalogicon*, ed. J.B. Hall (Turnholt, 1991).
—. *The Metalogicon of John of Salisbury: A Twelfth-Century Defense of the Verbal and Logical Arts of the Trivium*, trans. D.D. McGarry (Berkley and Los Angeles, 1962).
Stirnemann, P. "Review of C. De Hamel Glossed Books of the Bible and the Origins of the Paris Booktrade", *Bulletin Monumental* 143 (1985) 362-367.
—. "Histoire tripartite: un inventaire des livres de Pierre Lombard, un exemplaire de ses Sentences et le destinataire du Psautier de Copenhague", in *Du copiste au collectionneur: mélanges d'histoire des textes et des bibliothèques en l'honneur d'André Vernet*, eds J-F. Geneat and D. Nebbiai-Dalla Guarda (Turnhout, 1998), 301-318.
—. "En quête de Sens", in *Quand la peinture était dans les livre: Mélanges en l'honneur de François Avril*, eds E. König, C. Zöhl, M. Hofmann (Turnhout, 2007), 303-311.
Suckale-Redlefsen, G. *Katalog der illuminierten Handschriften der Staatsbibliothek Bamberg: Die Handschriften des 12. Jahrhunderts* 2 (Wiesbaden, 1995).
Turcheck, J.P. "A Neglected Manuscript of Peter Lombard's Liber Sententiarum and Parisian Illumination of the Late Twelfth Century", *The Journal of the Walters Art Gallery* 44 (1986), 48-69.
Roger of Wendover, *Roger of Wendover's Flowers of History*, trans. J.A. Giles vol.2 (London, 1849).
Yates Thompson, H., et al., *A Descriptive Catalogue of the Second Series of Fifty Manuscripts in the Collection of Henry Yates Thompson* (Cambridge, 1902).

CHAPTER THREE

SAINTS' ENCOUNTERS WITH SECULAR RULERS IN THE WELSH SAINTS' LIVES IN THE VESPASIAN LEGENDARY: MIRACLES BETWEEN BELIEF AND RELIGIOUS POLITICS

LUCIANA MEINKING GUIMARÃES

Introduction

The term "hagiography" is used in mainly two senses: it refers to a literary corpus comprising a variety of literary genres in prose or in verse and it refers to a field of study concerned with questions such as the purposes and functions of hagiographical texts[1] in a specific historical

[1] Known types of hagiographical texts are, for example: saints' lives, hymns, *visiones*, reports of *inventiones* and *translationes*, sermons, eulogies, *exempla*, dialogues, and others. According to Marc Van Uytfanghe, who uses the concept of hagiographical discourse to refer to the features that unite hagiographical texts, relevant features for the analysis of saints' lives are: 1. the relationship of the text's protagonist to God or to the divine, which, in the end, helps to differentiate biographies of ordinary historical characters from those of "saints"; 2. the heroic attributes of the saints due to hagiographical texts' debt to classical and Biblical models of expected heroic attributes and behaviours; 3. the oral background of some of the stories incorporated in saints' lives and the temporal distance to the real time in which the protagonists lived; 4. the composition of the text in terms of its functionality. See Marc Van Uytfanghe, "Die Vita im Spannungsfeld von Legende, Biographik und Geschichte (mit Anwendung auf einen Abschnitt aus der Vita Amandi prima)", in *Historiographie im frühen Mittelalter*, eds. Anton Scharer and Georg Scheibelreiter (Wien, 1994), 203-04. The hagiographer's intention is that of edifying the reader/listener and of apologetically idealising his/her

period, their liturgical uses, their form and structure, and the historicity of the figures portrayed in the texts.[2] Here, I define saints' lives as those texts composed by religious men to emphasise the glorious deeds of exemplary witnesses to the Christian faith from as early as the second century BCE. These texts developed throughout time, producing internal rules and literary conventions that distinguish them from other hagiographical texts.

My corpus comprises fourteen eleventh- and twelfth-century Welsh saints' lives found in the manuscript of the British Library Cotton Vespasian A.XIV. This manuscript, dated to ca.1200, contains the most important Welsh legendary extant, with the lives of some of the most famous saints of Wales. These are: 1. The Life of St. Gwynllyw (f.13); 2. The Life of St. Cadog by Lifris (f.17); 3. The Life of St. Illtud (f.43v); 4. The Life of St. Teilo (f.52); 5. The Life of St. Dubricius (f.56); 6. The Life of St. David by Rhygyfarch (f.61); 7. The Life of St. Dubricius by Benedict of Gloucester (f.71), 8. The Life of St. Brynach (f.77); 9. The Life of St. Padarn (f.80v); 10. The Life of St. Clydog (f.84v); 11. The Life

protagonist(s). Also important for the analysis of saints' lives is the influence of the hagiographer's milieu on the end-product: the hagiographer writes for an intended audience whose tastes leave an imprint on the text. Early studies on hagiography, although in many aspects outdated, provide illuminating insights as to the structure and methods of hagiographical texts. For this, see Hippolyte Delehaye, *The Legends of the Saints* (Dublin, 1905, reprinted 1998) and René Aigrain *L'Hagiographie. Ses sources, ses méthodes, son histoire* (Poitiers, 1953). For successful methodological approaches to saints' lives and questions relating to the historicity, contextualisation and uses of miracle stories, see F. Lotter, "Methodisches zur Gewinnung historischer Erkentnisse aus hagiographischen Quellen", *Historische Zeitschrift* 229 (1979), 298-356, B. Ward, *Miracles and the Medieval Mind, Theory, Record and Event: 1000-1215* (London, 1982), R. Bratož, "Der 'heilige Mann' und seine Biographie (unter besonderer Berücksichtigung von: Eugippius, Leben des heiligen Severin)", in *Historiographie im frühen Mittelalter*, eds. Anton Scharer and Georg Scheibelreiter (Wien, 1994), 222-52, M. Van Uytfanghe, "L'hagiographie: un 'genre' chrétien ou antique tardif?", *Analecta Bollandiana* 111 (1993), 135-188 and K. Herbers and D. Bauer, *Hagiographie im Kontext. Wirkungsweisen und Möglichkeiten historischer Auswertung* (Stuttgart, 2000).

[2] G. Philippart, "Hagiographies et hagiographie, hagiologes et hagiologie: des mots et des concepts", *Hagiographica* 1 (1994), 1-2 and F. Dolbeau, "Les Hagiographes au travail: collecte et traitement des documents écrits (IXe–XIIe siècles)", in *Manuscrits Hagiographiques et Travail des Hagiographes*, ed. Martin Heinzelmann (Sigmaringen, 1992), 49-52.

of St. Cybi (f.86); 12. The Life of St. Tatheus (f.88v); 13. The Life of St. Carannog (f.93); 14. The Life of St. Cybi (f.94v).[3]

Most of the Vespasian saints' lives have been seen as Welsh propaganda that responds to the steady Norman encroachment into Welsh land, and they have consequently been analysed in light of their purposes and functions within that specific historical moment.[4] The fact that the

---

[3] The edition used for the lives of Saints Gwynllyw, Cadog, Illtud, Carannog, Padarn, Brynach and Cybi is: Arthur Wade-Evans, *Vitae Sanctorum Britanniae et Genealogiae* (Cardiff, 1944). The edition used for the Life of St. David is: John William James, *Rhigyfarch's Life of St. David* (Cardiff, 1967). For the Life of St. Teilo, see J. Loth, "La Vie de Saint Teliau d'après de le Livre de Llandaf", *Annales de Bretagne* 9 (1894), 277-86. Joseph Loth collated the text of the Book of Llandaff with that in the *Vespasian* legendary, giving all the variant readings in the latter. The edited texts and translations for the lives of Teilo, Dubricius and Clydog were taken and, when necessary, adapted from William Jenkins Rees, *The Liber Landavensis, Llyfr Teilo* (Llandovery, 1840). For the edited text of the Life of St. Tatheus, see Harold Idris Bell, *Vita Sancti Tathei and Buched Seint Y Katrin: Reedited, after Rees, from the MSS. in the British Museum* (Bangor, 1909). For a description of the manuscript and its origin, see S. Harris, "The Kalendar of the *Vitae Sanctorum Wallensium*", *Journal of the Historical Society of the Church in Wales* 3 (1953), 3-53 and K. Hughes, "British Museum MS. Cotton Vespasian A.xiv ('Vitae Sanctorum Wallensium'): Its Purpose and Provenance", in *Studies in the Early British Church*, eds. Nora Chadwick, Kathleen Hughes, Christopher Brooke and Kenneth Jackson (Cambridge, 1958), 183-200.

[4] W. Davies, "Property rights and property claims in Welsh «Vitae» of the eleventh century", in *Hagiographie, cultures et sociétés: IVe-XIIe siècles*, eds. Pierre Riché and Évelyne Patlagean (Paris, 1981), 515-33, R. Bartlett, "Cults of Irish, Scottish and Welsh Saints in Twelfth-Century England", in *Britain and Ireland 900-1300: Insular responses to Medieval European Change*, ed. Brendan Smith (Cambridge, 1999), 67-86 and N. Jones and M. Owen, "Twelfth-century Welsh Hagiography: the Gogynfeirdd Poems to Saints", in *Celtic Hagiography and Saints' Cults*, ed. Jane Cartwright (Cardiff, 2003), 45-76, O. Edwards, "Welsh Saints' Lives as Legendary Propaganda", *Oral Tradition* 23 (2008), 148-58. For the Welsh twelfth-century political and ecclesiastical contexts, taking into consideration the production of the saints' lives, see C. Brooke, "The Archbishops of St. David's, Llandaff and Caerleon-on-Usk", in *Studies in the Early British Church*, eds. Nora Chadwick, Kathleen Hughes, Christopher Brooke and Kenneth Jackson (Cambridge, 1958), 201-41, C. Brooke, "St. Peter of Gloucester and St. Cadoc of Llancarfan", in *Celt and Saxon: Studies in the Early British Border*, ed. Nora Chadwick (Cambridge, 1963), 258-332, W. Davies, "Liber Landavensis: its construction and credibility", *The English Historical Review*, 88 (1973), 335-51, W. Davies, "The Consecration of Bishops of Llandaff in the Tenth and Eleventh Centuries", *Bulletin of the Board of Celtic Studies*, 26 (1976), 53-73. For studies

Welsh production of saints' lives is relatively late when compared to Ireland, Anglo-Saxon England, and Brittany[5] has also contributed to this view of a straightforward ecclesiastical response to the Norman alienation of Welsh lands.[6] However, the surviving Welsh saints' lives often cover

---

on the Welsh saints, their churches and cults, see G. Doble, "Saint Cuby (Cybi), A Celtic Saint. Patron of Duloe and Tregony, Cornwall, and of Holyhead and Llangybi (Card. Carn. and Mon.)", *Cornish Saints Series* 22 (1929), 3-47, G. Doble, "S. Carantoc, Abbot and Confessor", *Cornish Saints Series* 14 (1932), 3-41, G. Doble, "S. Patern", *Cornish Saints Series* 43 (1941), 3-51, G. Doble, "Saint Teilo", in *Lives of the Welsh Saints*, ed. Daniel Simon Evans (Cardiff, 1942, reprinted 1971), 162-206, G. Doble, "Saint Dubricius", in *Lives of the Welsh Saints*, ed. Daniel Simon Evans (Cardiff, 1943, reprinted 1971), 56-87, G. Doble, "Saint Illtud", in *Lives of the Welsh Saints*, ed. Daniel Simon Evans (Cardiff, 1944, reprinted 1971), 88-145 and Sabine Baring-Gould and John Fisher, *Lives of the British Saints* (London, 1907).

[5] For the sake of comparison, the earliest Breton hagiographical text, an anonymous life of Saint Samson, probably dates back to the seventh century, whereas the earliest extant Welsh hagiographical texts, the lives of Saint David by Rhigyfarch and of Saint Cadog by Lifris, were written at the end of the eleventh century. See C. Loomis, "King Arthur and the Saints", *Speculum*, 8 (1933), 478-82, J. Tatlock, "The dates of the Arthurian Saints' Legends", *Speculum*, 14-3 (1939), 345-65, N. Chadwick, "Intellectual Life in West Wales in the Last Days of the Celtic Church", in *Studies in the Early British Church*, eds. Nora Chadwick, Kathleen Hughes, Christopher Brooke and Kenneth Jackson (Cambridge, 1958), 124, W. Davies, "Property Rights and Property Claims in Welsh 'Vitae' of the Eleventh Century," in *Hagiographie, Cultures et Sociétés: IVe-XIIe siècles*, eds. Pierre Riché and Évelyne Patlagean (Paris, 1981), 515, E. Henken, "The Saint as a Secular Ruler: Aspects of Welsh Hagiography", *Folklore* 98 (1987b), 226-232, David Simon Evans, *The Welsh Life of St. David* (Cardiff, 1988), xxxix, F. Kerlouégan, "Les Pays Celtiques de IIIe au IXe Siècle", *Connaissance des Prères de l'Eglise* 49 (1993), 11 and Oliver Padel, *Arthur in Medieval Welsh Literature* (Cardiff, 2000), 37.

[6] The dearth of hagiographical material in Wales has been explained in at least two ways: firstly, in comparison to Ireland, where written material survives in manuscript collections (as opposed to the situation in Wales), Nora Chadwick suggests that the Irish centres of learning very soon found secular patronage of wealthy kings or chiefs who strongly supported intellectual production—see Chadwick, "Intellectual Life", 127. Another explanation is the loss of medieval manuscripts due to political instability or due to damage over time. This is perhaps the most probable scenario to be considered in the case of Wales. If important monastic libraries were pillaged in the costal areas of Anglo-Saxon England, the situation might also have been very similar in Ireland and Wales. The *Annales Cambriae*, for example, firstly attest to the presence of "the gentiles" among the

distinctive themes, like the veneration of bells and staffs, the taming of oxen and deer, and the saints' cursing and blinding of kings and church violators. This suggests that the eleventh and twelfth centuries witnessed a continuation of a Welsh hagiographical tradition that predates the extant manuscripts.[7]

Previous analyses of the production of the Welsh saints' lives have ignored their hagiographical and literary value while illuminating the political and ecclesiastical situation of Wales at the end of the eleventh and the beginning of the twelfth centuries. Scholars have often criticised the Welsh saints' lives for their "repetitive" character, especially in their suggestions that the miraculous accounts of these texts were uncreative and unoriginal. Such a view disregards the highly typological character of saintly biographies, which demands that the reader or listener take an allusive approach to Christian history and to the lives of individual Christians. This explains the inclination and predisposition to generalisation and de-personalisation in hagiographical narratives without prejudicing their contents to modern tastes.[8]

Miracles are the most conventional and predictable elements of saints' lives. They serve to emphasise the continuities and similarities within the Christian understanding of history. The *Vespasian Lives* conform to this typological understanding of sacred history, thus reproducing conventional miracles and Biblical motifs in a time when the hagiographical discourse[9] itself was being re-evaluated as part of the processes of canonisation and the struggles for metropolitan recognition on the part of many religious

---

Irish in the entry to the year 796. Viking activities also appear in the entries for the years 850, 853, and 895—when the Vikings apparently devastated areas as far as Gwent in southwest Wales. Be that as it may, there are references in medieval Welsh writings to hagiographical texts that served as sources for hagiographers and historians alike. For some examples, see Davies, "Property rights", 518. Chadwick, "Intellectual Life", 159-160, and Doble, "St. Teilo", 122-123.

[7] Davies, "Property rights", 518. This distinctive character approximates the hagiographical material of Wales, Ireland and Brittany, whose choice of themes and motifs is, to a certain degree, very similar. See Charles Plummer, *Vitae Sanctorum Hiberniae* (Oxford, 1910), J. Smith, "Oral and Written: Saints, Miracles, and Relics in Brittany, c. 850-1250", *Speculum* 65 (1990), 309-343 and Dorothy Ann Bray, *A List of Motifs in the Lives of the Early Irish Saints* (Helsinki, 1992).

[8] Earl, J., "Typology and Iconographic Style in Early Medieval Hagiography", *Studies in the Literary Imagination* 8 (1975), 15-46.

[9] See n.1 above.

communities throughout western Europe.[10] Having said this, my main objective is to address the question of the uses of miraculous accounts by twelfth-century Welsh hagiographers in the episodes in which encounters between the saints and secular persons, especially rulers, are described.

Most medieval saints' lives are divided into two sections: the *in-vita* and post-mortem sections. In the *in-vita* sections of the Welsh lives, we can uncover two types of miracles: those that function as a form of punishment based on Biblical models, and those that emphasise the powers of the saint over nature, based on secular tales of a localised character. The miracles in the post-mortem sections of the saints' lives deviate from these patterns, and suggest a changing attitude towards miraculous accounts and saintly interventions within the same legendary. The use of exorcism miracles, the references to relics brought up to the church and the references to contemporary historical events could point to a tendency on the part of twelfth-century Welsh hagiographers to adapt their accounts of historical events to a different understanding of Christian sacred history and, most importantly, to adapt their narratives to the current hagiographical trends in neighbouring English Benedictine monasteries.

Of the fourteen lives in Vespasian, two possess elaborate post-mortem sections: the lives of Saint Gwynllyw and Illtud. Of these, the post-mortem section in the life of Illtud seems to have been a later addition to an original text; the same seems to be true for the post-mortem section of the earliest life of Cadog. The two lives of Saint Dubricius possess post-mortem sections which are perfectly in line with twelfth-century cultic propaganda, with the miracles taking place at the saint's tomb. The life of Clydog, whose martyrdom takes place in the first lines of the life, narrates one posthumous miracle without a separate post-mortem section. The

[10] Ward, *Miracles*, 20-32. Martin Heinzelmann and Karl Herbers emphasised the fact that the production of miracle stories, from the early middle ages onwards, gradually distanced itself from a theoretical conceptualisation of miracles to more practical reasons connected to the promotion of a saint's cult. This reached its peak during the twelfth century. See M. Heinzelmann, "Die Funktion des Wunders in der spätantiken und frühmittelalterlichen Historiographie" in *Mirakel im Mittelalter: Konzeptionen, Erscheinungsformen, Deutungen*, eds. Martin Heinzelmann, Klaus Herbers and Dieter Bauer (Stuttgart, 2002), 19. For changes in the understanding and acceptance of miracles in relation to the canonization processes from the twelfth century onwards, see Michael Goodich, *Miracles and Wonders: the Development of the Concept of Miracle, 1150-1350.* (Hampshire, 2007).

remaining saints' lives show no attempt to list the saints' posthumous miracles. They show, on the other hand, an emphatic use of miracles in the episodes in which secular and religious values are confronted, especially in the encounters between a saint and a secular character. I will refer to these as "encounter scenes" or "encounter episodes".

Encounters between the Saints and Secular Characters in the Welsh Saints' Lives

The encounters of the saintly protagonists with secular characters in the Welsh saints' lives belong to one kind of episode slotted within these texts' underlying biographical pattern, which is similar to the pattern of secular heroic stories and recounts the main stages of the life of the saint, such as his/her birth, childhood and education. The saint is seen therefore as the main hero of the text.[11] By proving himself/herself in combat or in conflict with a secular ruler, the saint incorporates attributes equivalent to those of the folk-hero, with similar consequences in both cases: acquisition of land and of the rights to rule a specific territory.

---

[11] This has been demonstrated by Elissa Henken's study of the biographical patterning in Welsh hagiography. Henken demonstrated how the hagiographers, in describing their protagonists, moulded them in the fashion of their folk heroes. According to her, the pattern into which the Welsh saints' lives were modelled seemed to relate to the *rites de passage*, the most important life events in the life of a hero which normally tend to attract popular interest, fostering, therefore, the development of narrative accounts. She lists the following major stages of a Welsh saint's life: 1. Conception and birth; 2. Childhood (education); 3. Performing a miracle which indicates spiritual maturity; 4. Going out into the world—founding churches, making pilgrimages, retiring to the wilderness, journeying as a missionary; 5. Conflict with secular powers; 6. Ruling a territory; 7. Death. See Elissa Henken, *Welsh Saints: A Study in Patterned Lives* (Cambridge, 1991), 2. See also her earlier studies: "The Saint as Folk Hero: Bibliographical Patterning in Welsh Hagiography", in *Celtic Folklore and Christianity*, ed. Patrick Ford (Santa Barbara, 1983), 58-74, *Traditions of the Welsh Saints* (Cambridge, 1987a), "The Saint as a Secular Ruler", 226-232. The same association between saintly biographies and the heroic biographical pattern found in heroic folktales has been analysed for the Irish and Latin *Lives* of Irish Saints in Bray, *A List of Motifs*, 13-15. For motifs in Western hagiography in general, see R. Boyer, "An Attempt to Define the Typology of Medieval Hagiography", in *Hagiography and Medieval Literature: A Symposium*, eds. Hans Bekker-Nielsen and Peter Foote (Odense, 1981), 27-36.

All the direct encounters between secular people and the saints in these lives are linked, to different degrees, to miraculous accounts. It is normally after the saint's performance of a miraculous deed that the king and his retinue are made to repent for their sins. Miracles were by definition shared experiences within a community and it was through them that different cultural discourses are exchanged in the hagiographical texts:[12] hagiographers made their saints emulate miracles modelled on well-known miracles from the Old Testament and on the life of Christ described by the Evangelists. Also, they adapted and incorporated the "wondrous" elements found in the local secular or religious legends into the hagiographical material. Miracles stories constituted, apparently, part of the repertoire of a professional hagiographer[13] and their constant and reiterative appearance served not only to form the image of the saint, as suggested by Dorothy Ann Bray in her study of Irish saints' lives,[14] but also the image of secular rulers. This constant repetition is not only typologically conditioned but is also part of a textual strategy to symbolically accentuate the kind of social conflict that was an important issue to eleventh- and twelfth-century Welsh hagiographers, emphasising the virtues of the saints as protectors of the interests of their communities. This is how, to recall James Earl's words, saints' lives attempt to "draw *moralia* into the typological structures of sacred history."[15]

Encounters between saints and secular characters in the *in vita* sections of the Welsh *Lives*

"Typical" examples of encounter scenes in the *in vita* sections follow a structure in which either the saint settles in the territory of a specific secular ruler without his consent, provoking the ruler's rage, or in which the secular ruler or a secular person, mainly a robber or member of the king's retinue, invades a region under the saint's protection. Only specific types of miracles are used in the *in vita* sections of the Welsh lives. These have a tendency that allows their division into two main groups. The first

---

[12] Dieter Von der Nahmer, *Die Lateinische Heiligenvita* (Darmstadt, 1994), 146-52, Goodich, *Miracles and Wonders,* 5-34.
[13] The idea was suggested by D. Bray, "Miracles and Wonders in the Composition of the Lives of the Early Irish Saints", in *Celtic Hagiography and Saints' Cults*, ed. Jane Cartwright (Cardiff, 2003), 143.
[14] Ibid.
[15] Earl, "Typology", 16.

group includes the most pervasive miracles in the *Vespasian* legendary, and they convey the central idea of punishment administered by God for those who act against His plans and the agency of His saints. About 70% of the miracles belong to this group.

The commonest of these miracles are blindness and immediate destruction or immediate death. Examples of the latter are found in episodes in which rulers or their servants melt like wax, vanish like smoke, are destroyed by fire falling from heaven or are swallowed by the earth.[16] These miracles emulate Biblical models of punishment found especially in the Old Testament. Examples of punishment by blindness recall the fate of some Israelites in the Book of Isaiah due to hypocrisy and stubbornness, which prevented them from recognising God's message and His representatives.[17] The emulation of the earth's swallowing of Dathan and Abiram in Numbers also places the ruler in the saint's life in the role of those who acted against God's plans.[18] Dathan and Abiram's fate, together with destruction by melting like wax, reappears in Psalms.[19] Moreover, a destructive and punishing fire falling from heaven is also the fate of 250 Israelites who perish after both Dathan and Abiram's death.[20] These episodes in the Welsh saints' lives are often accompanied by paraphrases or direct quotations of the corresponding illustrative Biblical passages.

---

[16] The frequency and the types of miracles occurring in each of the lives is: blindness, Cadog (3x), Illtud, Padarn, and Cybi (1x each) and immediate destruction/immediate death: Cadog (1x fire falling from heaven, 1x person vanishes like smoke, 2x person swallowed by the earth), Illtud (3x person swallowed by the earth, 1x melting like wax), Teilo (1x unexplained sudden death), David (1x fire falling from heaven), and Padarn (1x unexplained sudden death, 1x person swallowed by the earth).

[17] See Isaiah 42:18-23. Here blindness is the main characteristic of a people who does not recognise God's message or His messengers. The same idea can be found in the Gospels, however unrelated to any of Christ's healings, but rather related to Christ's reprehension of the scribes and Pharisees for their hypocrisy and iniquity. See Mathew 23:13-29.

[18] The motif also occurs in a number of lives of Irish Saints—see Bray, *A List of Motifs*, 18 and D. Bray, "Saint Brigit and the Fire from Heaven", *Études Celtiques* 29 (1992), 105-113.

[19] The story of Dathan and Abiram's fate is told in Numbers 16:31 and in Psalm 105. Melting like wax is also mentioned in Isaiah 64:1-2.

[20] The miracle is also told in 2 Kings 1:1-14. In this episode, the prophet Elias makes fire come down from heaven upon two captains and their companions as a punishment for the insult they gave to the religion by consulting with other gods.

Two short passages from the life of Cadog may suffice to illustrate this kind of punitive miracle and the way in which the hagiographer deals with Biblical allusions. In some interrelated episodes of this life, Cadog meets Maelgwn, king of Gwynedd, and his son Rhun. These encounters happen, respectively, in chapters 23 and 24 of Wade-Evans' edition.[21] In an attempt to make peace with the king, Cadog causes a "column of mist" to go before him and descend over the king's encampment. An instance of temporary blindness follows:

> Mane autem facto, ab oratione surrexit, et ecce *columpna nubis precedebat eum*, que tentoria cuncta quoque agmina prescripti regis ope[r]iens obscurauit, et facta est dies ueluti tetra nox apud illus, ita ut nullus alterum contemplari ualeret. Tum in medio caliginis almus uir ante regis papilionem apparuit, salutans eum et ait, "Salue, rex. Queso, si in oculis tuis gratiam inueni, uerba mea benigne percipe".[22]

Chapter 24 describes the similar fate of Maelgwn's son Rhun in his encounter with Cadog. Rhun's men set Cadog's barn on fire and the smoke covers the whole region, reaching Rhun's encampment and all his army, who are consequently blinded:

> In illius plane hore momento, dum prelocutus Run in tabernaculo ludens in alea cum suis eunuchis consedisset, *fumus ad instar lignei postis* de sancti Cadoci horreo procedens, recto tramite per mane se ad ipsius papilionem tetendit, *lumenque occulorum omnium ibidem commanentium obcecauit.*[23]

In both episodes, a cloud or smoke that temporarily blinds the rulers is the factor responsible for their recognition of Cadog's saintly powers.

---

[21] See n.3 above.

[22] "When it was morning, he rose from prayer, and behold, *a column of mist went before him*, which, covering all the tents and the troops of the aforesaid king, beclouded them, and the day became to them as black as night, so that none was able to observe another. Then in the midst of the darkness the holy man appeared before the tent of the king, and saluting him says, 'Good health, o king. I beseech thee, if I have found favour in thine eyes, kindly to hear my words'" (Wade-Evans, *Vitae Sanctorum*, 74-5, my emphasis).

[23] "In the moment of that very hour, whilst the aforesaid Rhun was sitting in his tent playing at dice with his eunuchs, *the smoke like unto a wooden post* proceeding from the barn of saint Cadog, stretched itself right across throughout the morning to his tent, and *darkened the light of the eyes of all who were present in that place*" (Wade-Evans, *Vitae Sanctorum*, 76-7, my emphasis).

Similarly, the *columpna nubis* recalls the presence of God as the "pillar of cloud" in Exodus (Ex 13: 21): "Dominus autem *præcedebat* eos ad ostendendam viam per diem in *columna nubis*, et per noctem in columna ignis: ut dux esset itineris utroque tempore".[24] The cloud/smoke image is a punitive element that does not move before the saint but which arises on its own to punish Rhun's men. This recalls God's cloud against the pursuing Egyptians, which positioned itself between them and the Israelites.[25]

Although hagiographers frequently speak of wicked rulers, servants and/or robbers, none of these punitive miracles attribute their wickedness to the intervention of the devil, an attribution that only appears in connection to sin by sexual intercourse or in a single case of exorcism. Interestingly, the lives in which these two cases occur—those of Dubricius and Clydog—are strongly connected with the struggles of the see of Llandaff with St. David's and Hereford before the Roman curia for the control of the churches in which those saints were venerated. Both lives attempt to actualise their miracle stories in order to conform to contemporary miracle stories found in saints' lives written in continental and English Benedictine institutions. Due to its geographical location and its contact with English monasteries, Llandaff was from the outset an important institution to foster such actualisation of miraculous accounts in the hagiographical material written under her auspices.

The second group, on the other hand, includes miracles whose main function is to cause wonder and admiration on the part of the secular people who come into conflict with the saints.[26] The central idea conveyed

---

[24] "And the Lord went before them to show the way, by day in a pillar of a cloud, and by night in a pillar of fire; that he might be the guide of their journey at both times" (my emphasis).

[25] "Tollensque se angelus Dei, qui præcedebat castra Israël, abiit post eos: et cum eo pariter columna nubis, priora dimittens, post tergum stetit, inter castra Ægyptiorum et castra Israël: et erat nubes tenebrosa, et illuminans noctem, ita ut ad se invicem toto noctis tempore accedere non valerent", "And the angel of God, who went before the camp of Israel, removing, went behind them: and together with him the pillar of the cloud, leaving the forepart, stood behind, between the Egyptians' camp and the camp of Israel: and it was a dark cloud, and enlightening the night, so that they could not come at one another all the night" (Exodus 14:19-20).

[26] The frequency and the type of miracle occurring in each of the lives is: taming of wild and wondrous animals, Illtud (1x, stag), Cybi (1x, she-goat), Tatheus (1x, stag), and Carannog (1x, serpent); resuscitation of animals, David (1x, beasts), Brynach (1x, cow), and Tatheus (2x, 147 horses, 1 cow); changing the colour

by these miracle stories is that of the power of the saints towards men, and especially towards nature, with which the saints constantly interact. These episodes resemble secular stories and are often connected with an attempt to provide a topographical explanation for local and regional sites. The commonest miracles in this group are the taming of wild and wondrous animals; the resuscitation of animals (especially cows and horses); the change of colour and/or form of animals, human beings and/or objects; and the transformation of simple food into a marvellous banquet to be offered to a ruler. Of these, the taming of animals is the only "wonderful" act of a saint that does not necessarily relate to any attempt to explain a feature of the landscape; the animals simply serve to demonstrate the saints' powers over nature and to provoke awe and admiration.[27] It is possible to trace the origin of some of the miracles in this group to Biblical models, although the focus on popular admiration and wonder receives additional emphasis in these episodes.

One example of such an episode can be found in the life of Illtud. The eponymous saint invites a king called Meirchion to have breakfast with him in his cave. Meirchion does not find it suitable for a king to be offered water. Illtud's prayers make the water taste of several liquors and the king is satisfied:

> Potauit; potus sibi complacuit, et diuersorum liquorum sapores, uini et medonis inprimis, in aqua solummodo inuenit.[28]

The idea of making water taste like liquor is a variation of the Biblical motif of changing water into wine. It is unsurprising that the story mentions liquor instead of wine. In examples found in earlier Irish saints' lives, we find that the motif is malleable; in Cogitosus' life of St. Brigit, Cogitosus compares Brigit's turning water into ale with Christ's water-wine miracle at Cana.[29] The message behind this provision miracle is

---

and/or form of animals, human beings and/or objects, Cadog (2x, cows into bundles of fern, multi-coloured cows), Illtud (1x, men into stones), Brynach (1x, red-watered fountain), and Padarn (1x, men's souls in the form of ravens); transformation of simple food into a marvellous banquet, Illtud (2x, water and bread) and Brynach (2x, oak leaves and water).

[27] These types of miracles are also frequently found in medieval Irish lives—see Bray, *A List of Motifs*, 16-7, 89.

[28] "He drank; the drink pleased him, and the tastes of divers liquors, especially wine and mead, he found in water alone" (Wade-Evans, *Vitae Sanctorum*, 204-05).

[29] Bray, "Saint Brigit", 109-10.

symbolic: the provided water quenches the thirst for God's message,[30] the change in flavour emphasizes that the saint is God's elect.

Encounters between saints and secular characters in the post-mortem sections of the Welsh *Lives*

The hagiographiers seem to have changed the nature of the miracles they chose in the few post-mortem sections present in *Vespasian.* As mentioned above, three lives contain post-mortem sections: those of Gwynllyw, Illtud and Cadog. By thinking in terms of the legendary as a whole, however, and in terms of its production around 1200, we may notice subtleties embedded in the tales that narrate contemporary or near-contemporary matters.

In the life of Saint Gwynllyw, no direct encounter between the saint and a secular character is described in the *in vita* section. It is only in the post-mortem section that secular characters are incorporated as active characters and historical episodes relating to the Norman Conquest of England and Wales are interpreted. By taking sides with the Welsh and demonising the English rulers in their dealings with the Normans, the hagiographer of the life of Saint Gwynllyw engages in a rather historiographic activity, as it were, which finds parallels in the literary approaches employed by Norman historians when writing in their vernacular between the 1130s and the 1180s.[31] The nature of the miracles, with the hagiographer's emphasis on the act of telling or retelling of the miracles by eyewitnesses, demonstrates a careful attitude, as it were, towards miraculous interventions.

Here the miracles function as justifications for failed expeditions or invasions of land belonging to the community of St. Gwynllyw: all but two of king Gruffudd's ships sink in the middle of a storm, with the emphasis on the king's seeing the image of the saint sailing behind his fleet after the robbery. Harold Godwinesson loses the Battle of Hastings

[30] Jean-Pierre Charlier, *Signes et prodiges: les miracles dans l'Évangile* (Paris, 1987), 112-13.

[31] The historical works referred to are the works of Geoffrei Gaimar, *L'Estoire des Engleis*, Wace's *Le Roman de Rou* and Benoît de Sainte Maure's *Chronique des Ducs de Normandie*. For an analysis of the approaches adopted by these three writers and of their intended audiences, see P. Eley and P. Bennett, "The Battle of Hastings According to Gaimar, Wace and Benoît: Rhetoric and Politics", *Nottingham Medieval Studies* 43 (1999), 47-78.

because of his invasion of Gwynllyw's church. William Rufus and his soldiers almost die of hunger after returning from a punitive expedition in Glamorgan and violating the saint's church, even though, as the hagiographer affirms, the region was full of houses stocked with grain left by the fleeing population.

In the life of Illtud, the given cause for the death of the Anglo-Saxon king Edgar is his violation of the territories belonging to the community of St. Illtud. Here, the king also sees the figure of the saint in his dreams, who pierces him with his lance. The bell taken from Illtud's church by Edgar's men produces such a melodious sound that all the horses seem to be enchanted by it. Sparks of light appearing in the sky during a conflict that ended with the violation of Illtud's church are also interpreted as signs of the miraculous intervention of the saint. It is significant that, in most of these cases, the miraculous interventions are related to a vision, either in a dream or in a state of confusion, which makes personal the interpretation of the miraculous intervention.

The post-mortem section of the life of Cadog, on the other hand, which is a late insertion into the bulk of the text,[32] reveals a tension between an old model of miracle-accounts and a new one. In one and the same episode, the bellowing of the saint's shrine when struck by a violator is the explanation for a great earthquake in the region around Llancarfan. This corresponds again to an attempt to explain a natural phenomenon connected to the history of Cadog's church, although the violator is afterwards said to have melted like fire. Thus, the final punishment for the violation of the church follows a model from the Old Testament.

A tree is also made to bend itself under the feet of a preacher so that he is able to talk to a king who wanted to invade Glamorgan. There is one instance of a miracle taking place in front of the saint's altar, when iron rings bonding three strangers together break into pieces on the saint's festival, which approximates this miracle with those that were intended to foster the saint's cult at the site of his shrine and which are typical of twelfth-century cultic propaganda. Finally, there are two instances of resuscitation, one of an ox, the other of an overseer. All these miracles, apart from that of the iron rings, are retrospective, recounting episodes that happened in the lifetime of specific kings and not, as would be expected in post-mortem sections, recounting posthumous miracles associated with the figure of the saint. None of them is contemporary or near contemporary,

---

[32] See H. Emanuel, "An analysis of the composition of the 'Vita Cadoci", *The National Library of Wales Journal* 7 (1952), 217-227.

and here seems to lie the main difference between them and those miracles that provide an interpretative view of history, like in the lives of Gwynllyw and Illtud.

Finally, in the life of Dubricius, in the account of the *translatio* of the saint's relics from Bardsey Island to Llandaff under the auspices of Bishop Urban, for which purpose this life was written, a series of miracles take place before the eyes of all the present clergy and secular people when the relics are deposited in the church. The very specific location of the church as the site of the miracle *par excellence* is also characteristic of twelfth-century miracle collections. The emphasis in this kind of miraculous account is on the presence of reliable religious eyewitnesses who corroborate whether the events, be they healings or unexplainable facts, are miraculous or not.

Conclusion

It is not insignificant that the twelfth century witnessed a rise in the collection of posthumous miracles and the subsequent production of miracle collections to foster the cult of saints within the communities in which they were venerated. All throughout western Christendom, posthumous miracles were not only recorded to strengthen faith and convince non-believers of the intercession of particular saints, but also to promote their cults and function as advertisements of their shrines. In a context of conquest and colonisation, the uses of posthumous miracle collections or miracle lists are multifaceted and the influencing factors that moulded their final production are very difficult to grasp. The historical context of the encroaching Norman administration in Welsh monasteries was responsible for much of the production of saints' lives in south Wales during this period.

The Welsh hagiographers, by making use of a hagiographical genre with long-established traditional rules in their response to the Norman colonisation, succeeded in conforming to the contemporary hagiographical trend of western Europe. There are, however, some idiosyncrasies in the way that aspects of the lives of the Welsh saints are emphasised, especially in the way that their control over natural phenomena is emphasised. The Welsh lives present a very localised and regionalised exegesis of Biblical accounts in conflicts between religious and secular people. By differentiating miracles relating to punishment on the one hand and miracles representing the control of the saints' over natural phenomena on the other, the hagiographers demonstrate a very particular attitude in relation to Biblical miraculous events and to the popular attributions of wonderful events to saintly interventions. On the whole, however, there is a general conformity to well-established miraculous accounts found in the production of hagiographies in the western Christian world.

Finally, the post-mortem sections added to the main bulk of the text of a saint's life or produced in the second half of the twelfth century not only exemplify a kind of historical response to the steady conquest of Welsh ecclesiastical property by the Normans, but also demonstrate how this historical response was negotiated within the text. The emphasis is put on the interpretation of historical events as the result of the saints' interventions. The subtle aspect of a ruler's recognition of a miracle by his retelling of the events that he and his men observed, or by means of his very personal experience of the miracle, suggests a corroboration of a "historical" realm with a secular one. By putting emphasis on miracles taking place at the saint's tomb and at the much localised space of the saint's church, there is a change of focus within a single legendary in what regards a corroboration of miracles by a religious community. As such, the *Vespasian* legendary represents a kind of textual negotiation, as it were, between miracle stories following old hagiographical models and a new, rather continental and English model, in which posthumous miracles played an important role in ecclesiastical politics. These miracles had to conform to the changing understanding of miraculous interventions in the daily life of Christian communities.

Bibliography

Aigrain, René, *L'Hagiographie. Ses sources, ses méthodes, son histoire* (Poitiers, 1953).

Baring-Gould, Sabine and Fisher, John, *Lives of the British Saints* (London, 1907).

Bartlett, R., "Cults of Irish, Scottish and Welsh Saints in Twelfth-Century England", in *Britain and Ireland 900-1300. Insular Responses to Medieval European Change*, ed. Brendan Smith (Cambridge, 1999), 67-86.

Bell, Harold Idris, *Vita Sancti Tathei and Buched Seint Y Katrin, Reedited, after Rees, from the MSS. in the British Museum* (Bangor, 1909).

Boyer, R., "An Attempt to Define the Typology of Medieval Hagiography", in *Hagiography and Medieval Literature: A Symposium*, eds. Hans Bekker-Nielsen and Peter Foote (Odense, 1981), 27-36.

Bratož, R., "Der 'heilige Mann' und seine Biographie (unter besonderer Berücksichtigung von: Eugippius, Leben des heiligen Severin)", in

Historiographie im frühen Mittelalter, eds. Anton Scharer and Georg Scheibelreiter (Wien, 1994), 222-52.

Bray, D., "Miracles and Wonders in the Composition of the Lives of the Early Irish Saints", in *Celtic Hagiography and Saints' Cults*, ed. Jane Cartwright (Cardiff, 2003).

—. "Saint Brigit and the Fire from Heaven", *Études Celtiques* 29 (1992), 105-113.

Brooke, C., "The Archbishops of St. David's, Llandaff and Caerleon-on-Usk", in *Studies in the Early British Church*, eds. Nora Chadwick, Kathleen Hughes, Christopher Brooke and Kenneth Jackson (Cambridge, 1958), 201-41.

—. "St. Peter of Gloucester and St. Cadoc of Llancarfan", in *Celt and Saxon: Studies in the Early British Border*, ed. Nora Chadwick (Cambridge, 1963), 258-332.

Chadwick, N., "Intellectual Life in West Wales in the Last Days of the Celtic Church", in *Studies in the Early British Church*, eds. Nora Chadwick, Kathleen Hughes, Christopher Brooke and Kenneth Jackson (Cambridge, 1958), 121-82.

Charlier, Jean-Pierre, *Signes et prodiges: les miracles dans l'Évangile* (Paris, 1987).

Davies, W., "Liber Landavensis: Its Construction and Credibility", *English Historical Review* 88 (1973), 335-51.

—. "The Consecration of Bishops of Llandaff in the Tenth and Eleventh Centuries", *Bulletin of the Board of Celtic Studies* 26 (1976), 53-73.

—. "Property Rights and Property Claims in Welsh 'Vitae' of the Eleventh Century," in *Hagiographie, cultures et sociétés: IVe-XIIe siècles*, eds. Pierre Riché and Évelyne Patlagean (Paris, 1981), 515-33.

Delehaye, Hippolyte, *The Legends of the Saints* (Dublin, 1905, reprinted 1998).

Doble, G., "Saint Cuby (Cybi), A Celtic Saint. Patron of Duloe and Tregony, Cornwall, and of Holyhead and Llangybi (Card. Carn. and Mon.)", *Cornish Saints Series* 22 (1929), 3-47.

—. "S. Carantoc, Abbot and Confessor", *Cornish Saints Series* 14 (1932), 3-41.

—. "S. Patern", *Cornish Saints Series* 43 (1941), 3-51.

—. "Saint Teilo", in *Lives of the Welsh Saints*, ed. Daniel Simon Evans (Cardiff, 1942, reprinted 1971), 162-206.

—. "Saint Dubricius", in *Lives of the Welsh Saints*, ed. Daniel Simon Evans (Cardiff, 1943, reprinted 1971), 56-87.

—. "Saint Illtud", in *Lives of the Welsh Saints*, ed. Daniel Simon Evans (Cardiff, 1944, reprinted 1971), 88-145.

Dolbeau, F., "Les Hagiographes au travail: collecte et traitement des documents écrits (IXe–XIIe siècles)", in *Manuscrits Hagiographiques et Travail des Hagiographes*, ed. Martin Heinzelmann (Sigmaringen, 1992), 49-76.

Dorothy Ann Bray, *A List of Motifs in the Lives of the Early Irish Saints* (Helsinki, 1992).

Edwards O., "Welsh Saints' Lives as Legendary Propaganda", *Oral Tradition* 23 (2008), 148-58

Emanuel, H. "An Analysis of the Composition of the 'Vita Cadoci", *The National Library of Wales Journal* 7 (1952), 217-227.

Evans, David Simon., *The Welsh Life of St. David* (Cardiff, 1988).

Earl, J., "Typology and Iconographic style in early medieval hagiography", *Studies in the Literary Imagination* 8 (1975), 15-46.

Eley, P. and Bennett, P., "The Battle of Hastings According to Gaimar, Wace and Benoît: Rhetoric and Politics", *Nottingham Medieval Studies* 43 (1999), 47-78.

Goodich, Michael, *Miracles and Wonders: The Development of the Concept of Miracle, 1150-1350* (Hampshire, 2007).

Harris, S. "The Kalendar of the *Vitae Sanctorum Wallensium*", *Journal of the Historical Society of the Church in Wales* 3 (1953), 3-53.

Heinzelmann, M., "Die Funktion des Wunders in der spätantiken und frühmittelalterlichen Historiographie" in *Mirakel im Mittelalter: Konzeptionen, Erscheinungsformen, Deutungen*, eds. Martin Heinzelmann, Klaus Herbers and Dieter Bauer (Stuttgart, 2002), 23-61.

Henken, E. "The Saint as Folk Hero: Bibliographical Patterning in Welsh Hagiography", in *Celtic Folklore and Christianity*, ed. Patrick Ford (Santa Barbara, 1983), 58-74.

Henken, Elissa, *Traditions of the Welsh Saints* (Cambridge, 1987a)

Henken, E., "The Saint as a Secular Ruler: Aspects of Welsh Hagiography", *Folklore* 98 (1987b), 226-232.

Henken, Elissa, *Welsh Saints: A Study in Patterned Lives* (Cambridge, 1991).

Herbers, Klaus and Bauer, Dieter, *Hagiographie im Kontext. Wirkungsweisen und Möglichkeiten historischer Auswertung* (Stuttgart, 2000).

Hughes, K., "British Museum MS. Cotton Vespasian A.XIV ('Vitae Sanctorum Wallensium'): its Purpose and Provenance", in *Studies in the Early British Church*, eds. Nora Chadwick, Kathleen Hughes, Christopher Brooke and Kenneth Jackson (Cambridge, 1958), 183-200.

John William James, *Rhigyfarch's Life of St. David* (Cardiff, 1967).

Jones, N. and Owen, M., "Twelfth-century Welsh Hagiography: the Gogynfeirdd Poems to Saints", in *Celtic Hagiography and Saints' Cults*, ed. Jane Cartwright (Cardiff, 2003), 45-76.

Kerlouégan, F., "Les Pays Celtiques de IIIe au IXe Siècle", *Connaissance des Prères de l'Eglise* 49 (1993), 7-12.

Loomis, C., "King Arthur and the Saints", *Speculum* 8 (1933), 478-82.

Loth, J., "La Vie de Saint Teliau d'après de le Livre de Llandaf", *Annales de Bretagne* 9 (1894), 277-86.

Lotter, F., "Methodisches zur Gewinnung historischer Erkentnisse aus hagiographischen Quellen", *Historische Zeitschrift* 229 (1979), 298-356.

Padel, Oliver, *Arthur in Medieval Welsh literature* (Cardiff, 2000).

Philippart, G., "Hagiographies et hagiographie, hagiologes et hagiologie: des mots et des concepts", *Hagiographica* 1 (1994), 1-16.

Plummer, Charles., *Vitae Sanctorum Hiberniae* (Oxford, 1910).

Rees, William Jenkins, *The Liber Landavensis, Llyfr Teilo* (Llandovery, 1840).

Smith, J., "Oral and Written: Saints, Miracles, and Relics in Brittany, c. 850-1250", *Speculum* 65 (1990), 309-343.

Tatlock, J., "The dates of the Arthurian Saints' Legends", *Speculum* 14-3 (1939), 345-65.

Van Uytfanghe, M., "L'hagiographie: un "genre" chrétien ou antique tardif?", *Analecta Bollandiana* 111 (1993), 135-188.

—. "Die Vita im Spannungsfeld von Legende, Biographik und Geschichte (mit Anwendung auf einen Abschnitt aus der Vita Amandi prima)", in *Historiographie im frühen Mittelalter*, eds. Anton Scharer and Georg Scheibelreiter (Wien, 1994), 194-221.

Von der Nahmer, Dieter, *Die Lateinische Heiligenvita* (Darmstadt, 1994).

Wade-Evans, Arthur, *Vitae Sanctorum Britanniae et Genealogiae* (Cardiff, 1944).

Ward, Benedicta, *Miracles and the Medieval Mind: Theory, Record and Event: 1000-1215* (London, 1982).

CHAPTER FOUR

APOCALYPTIC TIME AND ANTI-SEMITISM IN THIRTEENTH-CENTURY ENGLAND

NANCY ROSS

Popular legend, contemporary events and medieval prophecy directed the interpretation of sacred scripture in the later middle ages. Before the twelfth century, scholars viewed the Biblical text of the Apocalypse as a description of the individual's internal battle with temptation. St. Augustine (ca.354-430), in his *De civitate Dei*, interpreted the Last Judgment of Revelation 20:12-13 to mean all divine judgments from the fall of Adam and Eve to the present. His interpretation of the Apocalypse was deeply personal and unrelated to historical events.[1] Exegetes held Augustine's view until the twelfth century, when Rupert of Deutz (ca.1075-1130) wrote his own commentary on the Apocalypse. Rupert broke with tradition and used the Apocalypse to explain historical events. Rupert's interpretation was well-received and several extant manuscripts were illustrated. A generation later, Joachim of Fiore (ca.1135-1202) drew parallels between the Old and New Testament eras in an effort to better understand and even predict what would happen in his own day. Preoccupied with finding patterns in history, Joachim identified particular historical events as representing specific visionary events in the Apocalypse. To these comparisons he added his own prophecy concerning the end of time, which he believed was imminent. Joachim of Fiore's views were popular in his own lifetime and led many to consider him a prophet. After his death, the Joachite movement spread quickly, emphasising the importance of Joachim's prophecies and their contemporary fulfilment.

The developments of Rupert and Joachim were, in part, anticipated by

[1] Barbara Nolan, *The Gothic Visionary Perspective* (Princeton, NJ, 1977), 4.

Berengaudus, the author of another commentary on the Apocalypse. This Berengaudus might be Berengaudus of Ferrièrs, a ninth-century monk, as argued by Derk Visser, or he might have lived as late as the end of the eleventh century in the east of France, south-east Belgium or the Rhineland, as argued by Nigel Morgan and Guy Lobrichon.[2] Whenever and where ever he lived, his Apocalypse commentary was of significant value, as there are numerous extant copies and records of others having existed in Benedictine libraries.[3]

Of the fifty or so extant illustrated Apocalypse manuscripts that were made in England ca.1250-1380, 15 include an abbreviated Berengaudus commentary. It normally appears in sections, paired with relevant segments of the the Apocalypse text and an appropriate illustration of the apocalyptic event. Two of these illuminated English manuscripts, the Gulbenkian and Abingdon Apocalypses, also include separate illustrations of the Berengaudus commentary.[4] These manuscripts were made in 1265-70 and 1270-75 respectively, with London being the mostly likely place of production.[5] The original patrons remain unknown, though they certainly moved within the court of Henry III.[6] Often thought to be a difficult text,

---

[2] Derk Visser, *Apocalypse as Utopian Expectation 800-1500, The Apocalypse Commentary of Berengaudus of Ferrieres and the Relationship between Exegesis, Liturgy and Iconography* (New York, 1996), 2-10, 87-103; Nigel Morgan, Suzanne Lewis, Aires Nascimento, Michelle Brown and Raquel Somoano, *Apocalypsis Gulbenkian* (Barcelona, 2002), 19, 46 n.9; Guy Lobrichon, "Une neouveauté: Les gloses de la Bible", in *Le Moyen Âge et la Bible*, eds Guy Lobrichon and Pierre Riché (Paris, 1984), 109; Guy Lobrichon, "L'ordre de ce temps et les désordres de la Fin: Apocalypse et société du IXe à la fin du XIIe siècle", in *The Use and Abuse of Eschatology in the Middle Ages*, eds Werner Verbeke, Daniel Verhelst and Andries Welkenhuysen (Leuven, 1988), 228.

[3] Morgan et al., *Apocalypsis Gulbenkian*, 19, 47 n.11.

[4] Lisbon, Gulbenkian Museum, MS L.A.139 and London, British Library Additional MS 42555. The Abingdon Apocalypse contains an Anglo-Norman French translation of the Berengaudus commentary.

[5] The artists of these two manuscripts made one other Apocalypse manuscript: London, Lambeth Palace Library MS 209. Morgan et al., *Apocalypsis Gulbenkian*, 13; Nigel Morgan, *Early Gothic Manuscripts 1250-1285, vol. 2* (London, 1988), 106-8.

[6] Ibid., 13. Suzanne Lewis argues that Giles de Bridgeport was the original owner of the Abingdon Apocalypse, but Nigel Morgan shows that the dates of possible patronage and production do not overlap. Morgan et al., *Apocalypsis Gulbenkian*, 36-7; Suzanne Lewis, "Giles de Bridgeport and the Abingdon Apocalypse", *England in the Thirteenth Century: Proceedings of the 1984 Harlaxton Symposium*

the inclusion of illustrations and commentaries with the Apocalypse text helped to make the visionary message more accessible to its aristocratic readers.[7]

The artists of the Gulbenkian and Abingdon Apocalypses placed the Apocalypse text and Berengaudus commentary below half page miniatures on opposite pages. The Gulbenkian Apocalypse has a total of 152 illustrations, with half of those illustrating the commentary. Such a sophisticated visual program likely involved the help of an ecclesiastic advisor, possibly one who was Franciscan.[8] In using this format, the artists separated and compared the two different realities presented by the texts. The Apocalypse illustrations reveal the divine, supernatural place of John's vision and the commentary illustrations depict life in the medieval world.

The physical world of the commentary illustrations specifically references thirteenth-century England and its challenges as perceived by the church. Berengaudus focuses his explanations on the problem of Jews in a Christian society, their crimes, and how society and the church should respond. The commentary illustrations are especially significant, as they add another layer of interpretation on top of Berengaudus and often take a stronger anti-Semitic position than what is found in his commentary text. More than half of the commentary illustrations include representations of Jews. Together, the exegesis and its illustrations establish a unique view of apocalyptic time and how it applies to Jews and their treatment by Christians in thirteenth-century England. The treatment of Jews in these two manuscripts, combined with new ideas regarding their place in apocalyptic time, contributed to an environment of urgency. Ultimately, these books served as propaganda for the expulsion of the Jews from England in 1290.

The presence of anti-Semitic themes in manuscript illumination is not new. Later medieval art is full of negative Jewish stereotypes, but the representations of the Jews in the commentary illustrations departed from conventional stereotypes and embraced newer, more worrying ideas. Evidence of this begins on f.4r of the Gulbenkian Apocalypse, where Ecclesia, the traditional personification of the church, is juxtaposed with

(1985), 107-119.

[7] It is possible that these commentaries were better known in the middle ages than the Apocalypse itself.

[8] Morgan et al., *Apocalypsis Gulbenkian*, 41.

Synagoga, who represents the Jews and their traditions.[9] Here, Synagoga drops the tablets containing the Mosaic Law, symbolically forfeiting her role as keeper of the Old Testament and passing it to the church. She is blindfolded, alluding to Judaism's inability to recognise the fulfilment of Old Testament prophecy in the figure of Jesus Christ, who sits next to Ecclesia in the form of a lamb. His blood spills into a golden chalice, reminding the reader of the Crucifixion and the sacrament of the Eucharist. While Synagoga sits on a mound of dirt in an antiquated structure representing the Temple of Solomon, Ecclesia is enthroned in a gothic church. She holds the banner of the Resurrection, a symbol of Christ's triumph over death, implying a connection between Judaism and spiritual death. This kind of image is common in late medieval art, but the inclusion of Old Testament prophets is unusual.

Most striking is the lack of unity in the reaction of these figures to the church. Only three of the four prophets look toward Ecclesia and affirm her position. The fourth prophet looks back to Synagoga, representing hesitation, though this hesitation is not discussed in the accompanying commentary. All of the prophets' heads are covered, in accordance with Jewish tradition, but the fourth prophet wears a *pileum*, the distinctive pointed hat worn by medieval Jews. The turning away by the fourth prophet reflects the medieval Christian belief in the stubbornness and ignorance of Jews in not accepting the fulfilment of their own prophecy in Jesus Christ. Not only does this image deride Judaism, but it also emphasises the wayward nature of its medieval adherents.

The commentary illustrations discuss the problem of relapsed Jewish converts to Christianity, accusations of ritual murder against the Jews, blasphemy, the problem of heresy, and fear of Antichrist. The treatment of Jews in the commentary cycle reveals the church's interest in the impact of the unconverted "chosen people" on the church and Christian society in general.

The visual polemic against the Jews in these two manuscripts is unique in that it ignores inherited crimes and focuses its attention on medieval crimes. Traditional anti-Semitic medieval imagery focuses on the belief that the Jews killed Christ. Medieval Jews were seen as inheriting the guilt of this crime and artists depicted Jews as Christ-killers, aiding the process of crucifixion. Traditional anti-Semitism also focuses on the greed of the Jews, who were often employed as moneylenders in medieval society. As a result, Jews were depicted in the visual arts and in literature as greedy

[9] All folio numbers refer to the Gulbenkian Apocalypse, unless stated otherwise.

usurers clutching moneybags and counting gold. These two themes constitute society's view of Judaism from antiquity throughout the middle ages, resulting in the Jews' unstable position within Christian society. While the commentary illustrations do depict Jews as usurers, they abstain from making references to Jews as Christ-killers. Instead, the commentary illustrations discuss the problem of relapsed Jewish converts to Christianity, accusations of ritual murder against the Jews, blasphemy, the problem of heresy, and fear of Antichrist and his associates.

In order to determine the nature of the anti-Semitism present in the Gulbenkian and Abingdon commentary illustrations, it is necessary to begin by establishing the iconography of the Jew. The *pileum* is the most common attribute of Jews in medieval art. Confusingly, not all Jews in the commentary illustrations wear *pilea*. Some hold moneybags, pickaxes, or magic wands. Others have distorted faces with absurdly large noses. This makes positive identification somewhat difficult. Sara Lipton discusses the matter in her analysis of anti-Semitism in the *Bible Moralisée*, early thirteenth-century French manuscripts that also illustrate Biblical and exegetical texts. In her book, she emphasises that standard Jewish attributes play a relatively small role in the iconography of Jews in these manuscripts, which is founded on visual uncertainty and suggestion. In her words, "The keynote in the iconography of Jews and Judaism in the *Bible Moralisée* is ambiguity".[10] Lipton's findings on this matter will be taken into account in considering the anti-Semitism present in the Gulbenkian and Abingdon commentary illustrations.

The Unconverting Jews

The conversion of the Jews to Christianity at the end of time was seen as a key eschatological event in the middle ages. Berengaudus' view of the conversion and its place in time explains the nature of the anti-Semitism in these manuscripts. He saw the apocalyptic time described in Revelation as representing earthly time. In his cxposition of Revelation 5:6 on f.6r, he interprets the seven horns of the lamb as the elect of the seven ages:

> By the seven horns all the elect are signified that are called to the kingdom of God, and this multitude we divide into seven parts. The first horn pertains to the elect who were before the Flood; the second to those who

[10] Sara Lipton, *Images of Intolerance: The Representation of Jews and Judaism in the "Bible moralisée"* (London, 1999), 15.

> were after the Flood to the time of the Law; the third to those who were under the Law; to the fourth the prophets; to the fifth the Jews who believed in Christ; to the sixth the people of the Gentiles; to the seventh those who are born at the end of the world and are to fight with Antichrist.[11]

The commentary illustration on the same folio describes the faithful of the seven ages in a visual time line, prefaced with images of the Resurrection and a congregation of priests praying before an altar. The first four time periods are represented by Adam and his descendants, Noah in the ark and the building of the Temple of Solomon. The fifth age is depicted with a group of Jews holding a banner reading "agnus dei". The sixth age shows the Baptism of Christ and the seventh and final age shows Antichrist preaching to his followers and a group of religious. Unlike the other ages, Berengaudus does not connect "the Jews who believed in Christ" with a particular Biblical or prophetic event, though he does place it in the past and emphasises this by following the image with the Baptism of Christ. The artist created a clear linear time line, leaving the reader with the clear impression that the prophecy concerning the conversion of the Jews was fulfilled during the life of Christ in the first century.

The Berengaudus commentary does not agree with the writings of the Tiburtine Sybil, which popularised the idea of an eventual Jewish conversion. The Sybil's writings were well-known in the later middle ages and adopted by Joachim of Fiore and others.[12] This late antique prophecy, written ca.380, assumed that the Jews have not yet been converted to the Christian faith. In anticipation of the event, the Sybil states that "When one hundred and twelve years have been completed, the Jews will be converted to the Lord."[13] The tone of the prophecy is hopeful and looks forward to the day when "Judah will be saved and Israel will dwell with confidence".[14] In contrast, the tone of the miniature is one of hopelessness concerning the Jews' future conversion.

The illustrative programme seeks to justify this anti-Semitic belief in the commentary illustration for Revelation 6:12-17 on f.13r. The exegesis

---

[11] Morgan et al., *Apocalypsis Gulbenkian*, 251.

[12] Bernard McGinn, *Visions of the End, Apocalyptic Traditions in the Middle Ages* (New York, 1979), 134-8.

[13] This prophecy comes from the Latin version of the Tiburtine Sybil. The period of one hundred and twelve years refers to the length of time that the king Constans will reign. McGinn, *Visions of the End*, 49.

[14] Ibid., 49.

interprets the opening of the sixth seal as

> the rejection of the Jews and to the calling of the Gentiles. And because Christ spoke many times in parables of the perdition of the Jews and to the choice of the Gentiles, he also did certain things which dignified this; he opened the sixth seal when he made clear that which he said and did in figures to the Doctors of the Church through the grace of his inspiration.[15]

In doing so, the author created a parallel with his interpretation of the horns of the lamb. Again, he informs the reader that the Jews have been rejected by God and that the reader is living during the turbulent period of the sixth seal.

In the accompanying image, the Jews turn their backs on the Veronica icon, which a gentile king touches with veiled hands, indicating his respect for the image. The commentary explains "By the earth in this case the Jews are signified: there was a great earthquake when this people was destroyed by the Romans."[16] Vespasian, the Roman Emperor, orders his agent to kill the Jews, which, in the image, happens as a result of their rejection of the image of Christ. Berengaudus and the artist both agree that this was an act of divine retribution, as the agent of the emperor looks to Christ as he stabs his victims. Instead of blessing the scene, Christ holds an episcopal seal, a symbol of a bishop's authority. The inclusion of the seal in this instance foreshadows a similar situation in thirteenth century England, where bishops had power over the Jews in their diocese. A green devil comes out of the mouth of one of the dying Jews, alluding to the frogs that come out of the mouths of the dragon, beast, and false prophet in Revelation 16:13, an obvious symbol of evil.[17] Christ's approval of the destruction of the Jews serves as a model of divine justice and rejection. There is no suggestion of optimism for future conversion.

Christ takes a more active role in the damning of the Jews in the commentary illustration on f.42v, showing the conversion of the gentiles and the damning of the Jews.[18] The artist depicts the faithful, who

---

[15] Morgan et al., *Apocalypsis Gulbenkian*, 262.

[16] Ibid., 263.

[17] The inclusion of a devil in this situation also foreshadows the commentary illustration to this verse on f.56v, which depicts devils coming out of the mouths of Jews.

[18] Morgan wrongly labels the illustration on f.42v as "The Conversion and Damnation of the Jews". Morgan et al., *Apocalypsis Gulbenkian*, 304. The reader could be led to believe that Jews are being converted, but the baptised figures do not hold any attribute that point toward their Jewish identity. Instead, a more

"constantly feel present in their hearts and ears the teaching of the saints in heaven" on the right hand of Christ.[19] The faithful obey the warning of the saints, present in the figure of St. John, and are baptised by him. Christ and an angel look down from heaven and bless them. On the left hand of Christ, a voice of thunder terrifies those who "are unable to ascend to perfection", interpreted by the artist as a group of Jews.[20] The thunder does more than inspire fear as it literally rains down upon the Jews, forcing them into the open jaws of Hell. A demon aids this process by pushing them into Hell with a stick. Meanwhile, the baptised faithful turn their backs on the Jews. To the right of the baptism, a group including a king, bishop and clerics face the Jews as they kneel in prayer. The artist's inclusion of this group and their positioning seem to indicate their involvement in the damning of the Jews, the object of their prayers.

Just as the faithful turn their backs on the Jews, the Jews turn their backs on Christian preachers. The commentary for Revelation 16:1-2 on f.52v interprets the voice from the temple to be the voice of the "Lord instructing preachers to announce the word of God to the elect but also to the reprobates, as far as the former can be reformed, so that the latter are condemned all the worse".[21] Berengaudus does not expect that the reprobates, a term referring to Jews in the thirteenth century, will be converted. Instead, he offers preaching as a means to condemn them. The left side of the miniature depicts an angel holding a scroll reading "*predicatoribus*", or "preachers", who points to a Franciscan friar. The Franciscan points toward the group of Jews, indicating that he is preaching to them. A Jew wearing a *pileum* emphatically rejects the message and turns his body away from the friar. Another turns away and a third debates with the kneeling preacher. Not one of the Jews is willing to accept the message, reflecting the contemporary belief that the Jews' refusal to convert to Christianity was due to their inherited stubbornness in misunderstanding their own scriptures. Instead of being converted by preaching, Berengaudus writes that the Jewish *reprobati* are "made worse by the preaching of saints", a statement that summarises the illustrative programme's attitude toward preaching to the Jews.[22]

The artist of the Abingdon Apocalypse replaces the friars of this image

---

accurate title would be "The Faithful and the Damnation of the Jews".

[19] Ibid., 304. Commentary for Revelation 14:2-5.

[20] Ibid., 304.

[21] Ibid., 317.

[22] Ibid., 317.

with a single bishop, emphasising the role that bishops also played in preaching. Beyond the replacement of the friars, these images suggest other differences in the perceived attitudes of the Jews toward the church and its authority. For example, the bishops' audience in the Abingdon miniature is much more attentive than the friars' counterpart. Although the Jew at the front of the group crosses his hands in defiance, all of the Jews in this group face the bishop, which is a more favourable response than what the friars receive. This could indicate the Jews' respect for episcopal authority, as they would have understood the way in which it extended to more secular affairs. English bishops in the thirteenth century frequently debated the economic and social issues regarding the Jews and had the power to limit their trade, movement and interaction with Christians in their diocese.

The established monastic orders in England had little to do with the Jews, but the newly constituted mendicant orders, arriving in England in the 1220s, took upon themselves the mission of preaching to the Jews. The friars usually settled near the Jewish quarter of a town, and were given permission to preach regularly to the Jews in their synagogues.[23] Several bishops were well known for their active anti-Semitism. Robert Grosseteste, Bishop of Lincoln (d.1253), advocated the excommunication of all Christians who associated with Jews or even sold them food.[24] John Peckham (ca.1225-92), the Franciscan Archbishop of Canterbury, devoted much of his time to compiling lists of Jewish abuses against Christians. He wrote letters to other religious and secular authorities lobbying for the restriction of their freedom.[25] Although the commentary images do not refer to specific people, they outline the role that the mendicant orders and bishops played in the struggle against the Jews.

As the commentary cycle progresses, the initial concepts of preaching and warning evolve into a physical battle between the preachers and the Jews. The artists convert the active apostolic mission of the Franciscans into a violent pictorial metaphor. In the miniature depicting Antichrist persecuting the church on f.58v, a friar holds a crucifix above his head, about to strike Antichrist, who wields a great sword. Ecclesia and the Whore of Babylon sit outside of the main action and their arrangement

---

[23] The Dominicans in particular were noted for establishing their houses in or near a Jewish settlement. Robert Stacey, "The Conversion of Jews to Christianity in the Thirteenth Century", *Speculum* 67 (1992), 263-83.

[24] Cecil Roth, *The Jews of Medieval Oxford* (Oxford, 1951), 21.

[25] Jeremy Cohen, *The Friars and the Jews*, (London, 1982), 80.

echoes the earlier comparison of the Ecclesia and Synagoga. In doing this, the artists draw parallels between Judaism and the Whore, who is one of the great destructive forces in the Apocalypse. Like the Jews in the commentary cycle, she turns away from Ecclesia. Her attention is focussed on the orb that she is holding, possibly alluding to the stereotype of the greedy Jew. A Hell-mouth supports the throne on which she sits. To further emphasise the connection between the Whore and Judaism, an angel points to a demon wearing a *pileum*.

Interpreted in this way, the fall of Babylon, described in Revelation 14:8 on f.61r, signifies the fall of Judaism. Berengaudus reinterprets the apocalyptic timeline in order to supplant traditional beliefs of Jewish conversion at the end of time. The program of commentary illustrations go beyond this and put forward the idea that just as God punished the Jews for their disbelief in ancient times, through Vespasian, so will he use the church, represented in the figures of Franciscans and bishops, to destroy the Jews at the end of time, embodied in the figure of the Whore of Babylon.

The Society of the Wicked

The commentary illustrations depict Jews as transmitters of a variety of evils, including usury, homosexuality, heresy, and ritual murder. Stories concerning these crimes spread quickly throughout Europe, often resulting in local and national campaigns against them. The church saw these evils as a mere glimpse of Jewish atrocities to come. Greater and more terrifying still, many believed that the thirteenth century would hail the arrival of the Antichrist, the dreaded Jewish Messiah.[26]

The commentary asks "In what manner will adore the dragon that is the Devil whom they do not see? … they will adore Antichrist… and in Antichrist the Devil".[27] The visual references to Antichrist in the commentary illustrations exceed the numerous references in the commentary text. The addition of commentary illustrations provides for his constant appearance throughout the Gulbenkian and Abingdon cycles. In the Gulbenkian Apocalypse, Antichrist appears fifteen times, far exceeding the visual treatment of this figure in any previous Apocalypse manuscript. The artists created visual parallels between Antichrist and the devil, showing

[26] Andrew Gow, *The Red Jews: Anti-Semitism in an Apocalyptic Age 1200-1600* (New York, 1995), 94.
[27] This excerpt is on f.35v. Morgan et al., *Apocalypsis Gulbenkian*, 294-295.

them as one and the same. The devil, in the guise of Abbadon, first appears in the miniature illustrating Revelation 9:1-12 on f.20r, depicting the unlocking of the abyss and the release of the locust beasts. Revelation 9:11 describes Abbadon as a king and the illustration shows him holding a sceptre, wearing a crown and sitting on top of one of the beasts. Although several of his attributes are similar to those held by divine creatures, such as his cruciform staff and extra sets of eyes on his wings, his devilish quality is revealed in his bat-like wings, grotesque facial features and monstrous followers.[28] Frightening and distorted, this supernatural being stands outside the experience of the reader. The realisation of the Abbadon's destructive potential requires a more human form, found in the figure of Antichrist.[29]

The Gulbenkian artists imposed another dimension on the link between the two figures, creating a connection not made in previous Apocalypse illustrations: Abbadon and Antichrist are represented as Jews. Abbadon's crown rests on top of a red *pileum* on f.20r.[30] The choice of colour makes this feature stand out against the gold background and dark blues of his cloak. By depicting him in this way, the artist ensured the reader's identification of the devil as a Jew. This same headdress reappears three more times on Antichrist. The artist's overt use of this feature makes Antichrist, Judaism and the devil inextricably linked in the mind of the reader. It also provided the reader with a framework within which to assess Jewish behaviour.

The connection between Antichrist and Judaism did not originate with the commentary illustrators, but with the Tiburtine Sybil. The text prophecies that Antichrist will be born of the tribe of Dan, one of the twelve tribes of Israel.[31] This apocryphal prophecy was validated through other Biblical prophecies concerning the tribe of Dan. Genesis 49 relates that Jacob, on his deathbed, blessed each of his twelve sons in turn, and

---

[28] The four apocalyptic beasts often contain extra sets of eyes on their wings, a reference to Ezekiel 1:6. Together with the cruciform staff, these holy attributes reveal Abbadon's deceptive qualities.

[29] The two figures are compared again in the miniature on f.26r, illustrating the execution of the two witnesses.

[30] The Hebrew word "Abbadon" appears in gold above his head. In Revelation 9:11, the Greek word "Appollyon" is also used, but the artist chose to use the Hebrew term, emphasising the devil's Jewish association.

[31] The Tiburtine Sybil was written in Greek ca.380 and was reworked and translated into Latin in the late tenth or early eleventh century. McGinn, *Visions of the End*, 43-44.

verse 17 says that Dan will "be a snake by the roadside, a serpent in the path, that biteth the horse's heels that his rider may fall backward". Dan's descendants are also excluded from the list of those numbered as the servants of God in Revelation 7. Although Biblical prophecy does not refer to Dan's connection with Antichrist, the author of the Tiburtine Sybil infers this, building upon canonical prophecy in order to create a believable figure of a Jewish Antichrist.

The artists of the Gulbenkian Apocalypse took great care to use Abbadon to introduce Antichrist, but once this was achieved, the focus shifted from Antichrist to his Jewish associates. Folio 34v depicts Antichrist speaking to his followers, who, with their *pilea*, are easily identified as Jews. His followers bring the seven vices, which, with the seven virtues, are personified on the right side of the miniature. This is a visual elaboration of the commentary, which foretells that Antichrist will rise "out of the society of the wicked".[32] The artists encouraged the reader to interpret "the society of the wicked" to mean "the society of the Jews". Other contemporary illustrations of Antichrist, such as those in the Lambeth Apocalypse, catalogue his deeds, but refrain from making the connection with the Jews.[33] In this illustration, and in the other commentary illustrations, Antichrist is portrayed as a source of conspiracy against the church, but not as the only conspirator; much of his sting comes from his influence over others.

As agents of Antichrist, the Jews are characterised as destroyers of the church. The image on f.19r references popular medieval legends concerning Jewish ritual murder. At the top of the miniature, Christ turns his attention from the apostles and points toward three registers of nefarious figures. Two Jews with swords commit murder in the upper register. Antichrist sits between them, wearing a crown of thorns in imitation of Christ.

This illustration would have reminded the reader of past ritual murder accusations against the Jews. St. William of Norwich (d.1144) and Little St. Hugh of Lincoln (d.1255), both children, were believed to have been abducted and murdered by Jews who were performing ritual sacrifices. The written accounts of these saints do not focus on the holiness of the children, unlike the accounts of many saints' lives, but emphasise their cruel martyrdom at the hands of the Jews and their posthumous miracles.

---

[32] Morgan et al., *Apocalypsis Gulbenkian*, 293.

[33] The Lambeth Apocalypse includes six illustrations in the bottom margins of ff.11v-13r illustrating scenes from the life of Antichrist.

Although the evidence in both cases did not actually point to the Jews as the perpetrators of these crimes, rumours of their guilt spread quickly and were readily believed by suspicious Christian communities.[34] Jewish communities and the church continually refuted the charges of Jewish ritual murder, but these efforts did not serve to lessen the tension between the two communities. By including visual references to these legends, the artist wanted to inflame the reader's resentment of the Jews.

In the middle register of the same miniature, two events are taking place. On the left, two men sit naked, one staring at the other holding his penis and representing the sin of homosexuality. On the right, two men count their money. The figure on the left holds a money bag, clearly identifying him as a Jew. This scene reflects the Jews' role as moneylenders. As Christians were forbidden to lend money with interest, this became the occupation of many Jews, who were barred by law from many other professions. They were, therefore, perceived as being greedy. These two scenes were appropriately placed together as medieval theology connected sexual sins with usury.[35]

In the bottom register, an open Hell-mouth holds embracing couples and churches. The figure on the far left wears a *pileum*. It is unclear if he is embracing a man, a further reference to Jewish homosexuality, or a woman. This image could represent the dangers of Jews intermarrying with Christians, which was starting to happen in England during the thirteenth century. The artists show that the role of the Jews in the destruction of the church is clear, as the physical structures of churches are among the damned in Hell.

The commentary illustration on f.43v shows a similar process, except here the devil literally breaks the church apart with a spear. In these two images, and throughout the Gulbenkian and Abingdon commentary cycles, the crimes of the Jews, and particularly their greed, are emphasised and connected with the potential downfall of the church. To the right, a group of Jewish moneylenders conduct their business. This is probably an allusion to the fact that many English monastic establishments borrowed money from Jews and used church property as security in the loans. The

---

[34] David Farmer, *The Oxford Dictionary of Saints*, 245.

[35] Usury was seen as the unnatural procreative act of money, as money gave birth to more money in the form of interest. This was associated with non-procreative sexual acts, which, whilst imitating heterosexual sex, did not produce children and were seen by the church as an offence to natural law. Anne Derbes and Mark Sandonna, "Barren Metal and the Fruitful Womb: The Program of Giotto's Arena Chapel in Padua", *Art Bulletin* 80 (1998), 274-291.

collective debt that monasticism owed Jewish lenders was known to be a sizeable amount. Jewish usury is portrayed as having a direct effect on the stability of the church. This may have been partially true, as serious debt would cripple any institution, but the responsibility of borrowers is not questioned and the blame is placed on the lenders, whose crimes are then augmented in order to prejudice the reader against them.

Probably the most surprising Jewish crime of this period was heresy. The commentary text on f.19r describes heretics as those "who attempted to attack the church of God in many ways".[36] In addition to the many ways that Jews have already been shown to attack the church, many clerics considered them to be heretics in that they denied the fulfilment of Old Testament prophecy in Jesus Christ. Their obvious rejection of Christian preachers was seen as an attack. Medieval theologians accused Jews of not understanding their own books, or, in some cases, of understanding Christian truths and then denying them.

"Who is the liar but he who denieth that Jesus is the Christ? He is Antichrist" (1 John 2:22). The argument put forward by the artists of these illustrations is clear: an involvement with the Jews constitutes an involvement with Antichrist and the devil. The medieval readers of these books were bombarded with written and pictorial messages that the Jews did not have a proper place in Christian society.

The Wandering Jews

The commentary illustrations include subtle references to several contemporary legends involving Jews. One particularly significant legend is that of the Wandering Jew, which was popularised in the thirteenth century. Visual references to this legend encouraged the reader to believe that God has predestined the Jews to wander. Within the illustrations, Jewish wanderers stand in opposition to the faithful wanderers and burden the church. When viewed within the context of the social tensions between Christians and Jews at the time, the illustrations offer a divinely approved solution for solving the perceived problem of English Jewry. These early visual suggestions of the Wandering Jew helped to promote the legend and offered a religious justification for the expulsion of Jews from England in 1290. The way in which the artists focus on the contemporary application of the legend and not on the role of the Wandering Jew in the Passion is similar to the treatment of other anti-Semitic themes in the commentary

---

[36] Morgan et al., *Apocalypsis Gulbenkian*, 272.

cycle.

Written reports of sightings of the Wandering Jew, beginning in the thirteenth century, were a crucial part of the dissemination and credibility of the legend. The origin of the story is unknown and probably existed in an oral form long before it appeared in writing.[37] The first English documentation of the legend appears in the writings of Roger of Wendover, a chronicler at the monastery at St. Albans.[38] In an entry for the year 1228 in his *Flores Historiarum*, Roger describes a conversation with an Armenian archbishop visiting St. Albans. During the course of the visitor's entertainment, the archbishop was asked if he had ever seen or heard of a man called Joseph, "a man of whom there was much talk in the world, who, when our Lord suffered, was present and spoke to Him, and who is still alive in evidence of the Christian faith".[39] The archbishop's knight, acting as an interpreter, replied "My lord well knows that man… the said Joseph ate at the table of my lord the archbishop of Armenia, and he was often seen and held converse with him".[40] He then related Joseph's story:

> At the time of the suffering of Jesus Christ, He was seized by the Jews and led into the Hall of Judgement, before Pilate the governor, that He might be judged by him on the accusation of the Jews, and Pilate finding no cause for adjudging Him to death, said to them: "Take Him and judge Him according to your law"… When therefore the Jews were dragging Jesus forth, and had reached the door, Cartaphilus, a porter of the hall in Pilate's service, as Jesus was going out of the door, impiously struck Him on the back with his hand, and said in mockery: "Go quicker, Jesus, go quicker; why do you loiter?" And Jesus looking back on him with a severe countenance said to him, "I am going, and you will wait till I return".[41]

According to legend, Cartaphilus was later baptised into the Christian faith by Ananias, who was baptised by Paul, and renamed Joseph. The knight said that Joseph wanders in the east, telling stories of the Resurrection and

---

[37] George Anderson, *The Legend of the Wandering Jew* (Providence, 1965), 11-20.

[38] The only extant version prior to Roger's was in a monastic chronicle of Ferraria, in southern Italy, in an entry for the year 1223. It seems unlikely that the early Italian version had any bearing on the English versions. R. Edelmann, "Ahasueros, The Wandering Jew, Origin and Background", *The Wandering Jew; Essays in the interpretation of a Christian Legend* (Bloomington, 1986), 1-10.

[39] Anderson, *The Legend of the Wandering Jew*, 18.

[40] Ibid., 18.

[41] Ibid., 18-19.

the apostles to large crowds, refusing "all gifts that are offered to him, being content with slight food and clothing".[42] This account does not specifically say that Cartaphilus was a Jew, but his association with the Jews was implied.

The story of this legend must have been widely known, as an image of it appears in the De Brailes Hours, made in Oxford at the independent workshop of William de Brailes, circa 1240.[43] Four medallions depicting scenes from the Passion make up the frontispiece to the office of Sext on f.43v. Raised hands and pointed fingers represent the exchange between Jesus and a man in the upper-left medallion. An inscription above the illustration reads "…regarde et dit e tu remiens ices desque ieo reveine", indicating that this man is Cartaphilus.

A few years later, Matthew Paris succeeded Roger of Wendover as the St. Albans chronicler. In his *Chronica majora*, Matthew retold the story of the Armenian archbishop in the entry for the year 1252.[44] A drawing on f.74v accompanies the written account and shares the iconography of the de Brailes image. The words in the scrolls, representing conversation, identify the hunched figure as Cartaphilus. This image combines the two scenes in the top register of the de Brailes frontispiece by depicting Cartaphilus as mocking Christ on the road to Calvary, instead of outside the judgement hall. Matthew Paris's fusion of these two events caused Cartaphilus' mocking to appear more malicious, as he is taunting a man condemned to death. This version retains the pointing gesture of the Jew, but adds a pickaxe. Although this might seem like an obscure attribute, it would have led the visually literate medieval reader to compare Cartaphilus with Cain, who was sentenced by God to be a "fugitive and a vagabond shalt thou be upon the earth" (Genesis 4:12) after killing his brother Abel. In medieval art, Cain sometimes carries a pickaxe as his attribute.[45] A letter from Pope Innocent III to the Archbishop of Sens and the Bishop of Paris, dated 15th of July, 1205, established the connection between the two figures. The pontiff makes the role of contemporary Jews clear in stating that

> Christian piety accepts the Jews, who, by their guilt, are consigned to perpetual servitude because they defied the Lord… like Cain, they are

---

[42] Ibid., 19.
[43] London, British Library Additional MS 49999.
[44] Cambridge, Corpus Christi College MS 16.
[45] Suzanne Lewis, "Tractatos Adversus Judeos", *Art Bulletin* 68 (1986), 543-65.

> [condemned] to be wanderers and thieves".[46]

Cartaphilus wears a pilgrim's hat around his shoulders, emphasising his sentence to wander. In both illustrations, in the written accounts, and in the papal letter, the themes of Jewish guilt and subsequent wandering are dominant.

The first suggestion of the legend in the Gulbenkian and Abingdon cycle reveals itself in a commentary illustration on f.3r, which depicts Christ as the Door of Heaven. A large and diverse group, divided into two sections, stand before Christ in Heaven. The people on the left include a king, bishops, and clerics, who all stand tall, while those on the right bend down. It appears as if this entire group is in the process of kneeling before Christ in an act of devotion, but this is not the case. The meaning of the image can be found in the commentary text as the bending figures are described as being "much weighed down with guilt".[47] A prominent bearded figure in the centre carries a pickaxe. It is possible that the artists were familiar with visual representations of the legend of the Wandering Jew and consciously redeployed it here. Indeed, they were familiar with an exceptionally wide range of anti-Semitic arguments and images. To the right of the figure in question, two men carry staves, indicating that they are wanderers. Several of the figures wear Phrygian caps, which some Jews in the Gulbenkian and Abingdon Apocalypses wear instead of *pilea*. The figure just below the foremost figure with a staff wears a green *pileum*, confirming his and the bending figures' Jewish identity.

This image highlights the larger issue addressed in the commentary, which reads

> For every one who enters through the door into the sheepfold that is the Holy Church will be saved: the reprobates whose number is countless… are included within the Holy Church, but by what right will they enter?[48]

The answer to this question lies in the illustration, which shows that the Jews are too heavily burdened by their own guilt to enter Heaven. The presence of the pickaxe represents the Jews' guilt as the mockers of Christ and Christianity. If the Jews are technically part of the church, as the commentary suggests, but do not accept its teachings and authority, then

---

[46] Ibid.

[47] This passage from the commentary interprets Revelation 4:1-9. Morgan et al., *Apocalypsis Gulbenkian*, 247.

[48] Morgan et al., *Apocalypsis Gulbenkian*, 247.

their existence in Christian society comes into question.

The illustrative programme makes one other reference to the legend. On f.40r, the miniature depicts the false prophet ordering his followers to be marked, found in Revelation 13:16-18. The figure to the left of the horseman fits physical description of the Wandering Jew.[49] This man wears little clothing and has bare feet, which confirm his poverty, and carries a staff, which shows that he is a traveller. Like the Wandering Jew of the Matthew Paris drawing, he wears a pilgrim's hat tied around his shoulders and his hand is raised in a gesture of mocking. Unlike the Jew in the legend, this man carries a money bag, signifying his affiliation with contemporary Jews. Although the Armenian archbishop claimed that Cartaphilus was baptised, the illustrator ignores this aspect of the legend by showing Cartaphilus' devotion to the false prophet. Instead of representing the legend specifically, this man represents the general premise of the legend, which holds that the Jew is cursed to wander until the return of Christ at the end of the world. His placement in this scene draws attention to his guilt in the legend. As the image depicts several of the false prophet's other followers as Jews in this image, and it was believed by Christians that the Jews would follow Antichrist, the artists of these manuscripts turn the legend of the Wandering Jew into the legend of the Wandering Jews. The emphasis on contemporary Jews in the commentary cycle leads the reader to view them as divinely cursed wanderers.

The Jewish wanderers are balanced in the commentary cycle by the inclusion of a number of saintly wanderers. Numerous bare-footed Franciscans and Dominicans appear as contemporary apostles, preaching to the faithful and the unconverted.[50] St. Christopher, the patron saint of travellers, appears on f.60v in connection with the faithful storing their treasures in heaven.[51] The legend of St. Christopher, which appears in many medieval English wall paintings, would have been well known to the artists. His bare foot, worn out shoe, and travelling cloak emphasise his wandering nature. Although he carries an infant, who carries the weight of

---

[49] This figure appears in the same scene in other Apocalypses. His presence is particularly notable in the Lambeth Apocalypse on f.20v, where he opens his mouth to speak, a feature that seems to replace the speech scrolls of Matthew Paris's Wandering Jew. Cartaphilus' words, represented in gesture or stated in scrolls, are probably his most important attribute, as seen in the original De Brailes illustration.

[50] Bare feet symbolise apostolic poverty.

[51] This image illustrates the commentary for Revelation 18:21-24.

the world on his back, St. Christopher stands tall, which reflects his diligence in serving God. His holy burden was in opposition to the Jews' burden of guilt on f.3r.

The most prominent saintly wanderer is St. Paul, who appears nine times in the commentary illustrations.[52] His presence is appropriate to the anti-Semitism in these manuscripts, as he represents the Jews who were converted at the time of Christ. Although he was not one of the original twelve apostles, he was seen as taking Judas' place in the group. St. Paul represents the ideal Jew, who not only converts to Christianity, but becomes a great asset to the church.

Paul usually appears in the commentary illustrations with the faithful in heaven and often sits opposite St. Peter. His attributes are his receding hairline and the sword. The most poignant image of Paul is on f.67v, which contrasts him with a Jew. The figure standing in the Hell-mouth is meant to represent Antichrist, as indicated in the commentary.[53] Throughout the commentary cycle, Antichrist usually holds a sword and wears a crown, but the figure in this image does not have either. Without them, Antichrist is depicted as an ordinary Jew wearing a *pileum*. He may even represent Jews generally, as he turns away from Christ and Paul and is lead into Hell by the devil. The commentary supports a general interpretation of the figure of Antichrist by saying that "others were slain… of he who sat on the horse; for their ruin he ordained their destruction".[54] By interpreting the "others" to mean the Jews in this illustration, the artist states visually that the Jews are destined, even divinely "ordained", for destruction, which takes the form of wandering and displacement in this cycle.

The presence of the legend of the Wandering Jew in the illustrations reveals the artists' manipulation of existing sources for the purpose of deploying a particular political agenda. The illustrations offered and justified theologically a solution to a perceived economic, social, and political problem. After the expulsion of the Jews in 1290, one chronicler wrote that the Jews had been "condemned to wander the earth and to be cut off".[55] The writer's allusion to the legend of the Wandering Jew echoes the same themes put forward by the illustrators.

---

[52] The only other saint, unconnected to the narrative of the Apocalypse, who appears more than Paul is Peter.

[53] This is the commentary excerpt for Revelation 19:20-21. Morgan et al., *Apocalypsis Gulbenkian*, 336.

[54] Ibid., 336.

[55] Robin Mundill, *England's Jewish Solution: Experiment and Exclusion, 1262-1290* (Cambridge, 1998), 1.

Although the Jews were not expelled from England until the end of the thirteenth century, tension between Jews and Christians in England had been growing for more than a hundred years.[56] The 1260s signalled a change in the way that the church and the state dealt with Jewry. During this period, a number of violent attacks, including Simon de Montfort's infamous sacking of the London Jewry in 1264, coincided with religious and political attacks on Jews.[57]

The commentary illustrations reflect the heightened tensions of the period, but the aggressive character of their anti-Semitism, found in the wide variety of arguments used against the Jews, encouraged and actively promoted this prejudice. Derk Visser describes the Berengaudus commentary as ultimately optimistic.[58] A degree of that is to be expected, as the Apocalypse itself is ultimately optimistic for Christians, ending with the descent of God's kingdom on earth. Together with its illustrations and legacy, though, the Berengaudus text is more sinister than optimistic, denying medieval Jews the opportunity of Christian salvation. By placing these arguments within an apocalyptic framework and by highlighting Berengaudus' unique views of comments on time, these manuscripts contributed to an environment of urgency in addressing these perceived problems. However perverse it might seem to the twenty-first century viewer, the commentary cycle provided the reader with a developed strategy for identifying, discussing and solving the alleged problems caused by contemporary Jews and helped to instigate a domestic crusade on English Jewry.

Bibliography

Anderson, George, K., *The Legend of the Wandering Jew* (Providence, 1965).

Chazan, Robert (ed), *Church, State, and Jew in the Middle Ages* (West Orange, 1980).

Cohen, Jeremy, *The Friars and the Jews: The Evolution of Medieval Anti-Judaism* (London, 1982).

Derbes, Anne and Mark Sandonna, "Barren Metal and the Fruitful Womb:

[56] Edward I announced the expulsion of the Jews from England in a decree dated November 5, 1290. Robert Chazan, ed, *Church, State, and Jew in the Middle Ages* (West Orange, NJ, 1980), 319.

[57] Mundill, *England's Jewish Solution*, xxv.

[58] Visser, *Apocalypse as Utopian Expectation*, 3, 22, 105-110, 115, 121, 127, 128.

The Program of Giotto's Arena Chapel in Padua", *Art Bulletin* 80 (1998), 274-91.

Edelmann, R., "Ahasuerus, The Wandering Jew", in *The Wandering Jew: Essays in the Interpretation of a Christian Legend*, eds Galit Hasan-Rokem and Alan Dundes (Bloomington,1986), 1-10.

Farmer, David, *The Oxford Dictionary of Saints* (Oxford, 1997).

Gow, Andrew Colin, *The Red Jews: Anti-Semitism in an Apocalyptic Age 1200-1600* (New York, 1995).

Lewis, Suzanne, "Giles de Bridgeport and the Abingdon Apocalypse", *England in the Thirteenth Century: Proceedings of the 1984 Harlaxton Symposium* 1 (1985), 107-119.

—. "Tractatos Adversus Judeos", *Art Bulletin* 68 (1986), 543-565.

Lipton, Sara, *Images of Intolerance: The Representation of Jews and Judaism in the "Bible moralisée"* (London, 1999).

Lobrichon, Guy, "Une neouveauté: Les gloses de la Bible", in *Le Moyen Âge et la Bible*, eds Guy Lobrichon and Pierre Riché (Paris, 1984), 95-114.

—. "L'ordre de ce temps et les désordres de la Fin: Apocalypse et société du IXe à la fin du XIIe siècle", in *The Use and Abuse of Eschatology in the Middle Ages*, eds Werner Verbeke, Daniel Verhelst and Andries Welkenhuysen (Leuven, 1988), 221-241.

McGinn, Bernard, *Visions of the End, Apocalyptic Traditions in the Middle Ages* (New York, 1979).

Morgan, Nigel, *Early Gothic Manuscripts 1250-1285* vol. 2 (London, 1988).

Morgan, Nigel, Suzanne Lewis, Aires Nascimento, Michelle Brown and Raquel Somoano, *Apocalypsis Gulbenkian* (Barcelona, 2002).

Mundill, Robin R., *England's Jewish Solution: Experiment and Expulsion, 1262-1290* (Cambridge, 1998).

Nolan, Barbara, *The Gothic Visionary Perspective* (Princeton, 1997).

Roth, Cecil, *The Jews of Medieval Oxford* (Oxford, 1951).

Stacey, Robert C., "The Conversion of Jews to Christianity in Thirteenth Century England", *Speculum* 67 (1992), 263-83.

Visser, Derk, *Apocalypse as Utopian Expectation 800-1500, The Apocalypse Commentary of Berengaudus of Ferrieres and the Relationship between Exegesis, Liturgy and Iconography* (New York, 1996).

CHAPTER FIVE

ART AND BELIEF IN MEDIEVAL CASTILE

TOM NICKSON

On a quiet dawn on the 22nd August 1936, the Dean of Toledo cathedral, don José Polo Benito, was shot by Republican forces outside the *Puerta del Cambrón* in Toledo.[1] The Toledan dean was one of countless victims of a civil war in which art, belief and the memory of Spain's medieval past were commandeered in the effort to demarcate boundaries between Nationalist and Republican forces.[2] Eleven years earlier Polo Benito, a distinguished writer and academic, had published an important booklet on the mural paintings of the St. Blaise chapel in the cloister of Toledo cathedral, in which he demonstrated that the selection of Christological scenes on the chapel's upper walls corresponded to the articles of the Creed.[3] The chapel was painted in the 1390s, less than thirty years after the conclusion of another bloody peninsular civil war in which, as in the 1930s, the victor had portrayed himself as a defender of orthodoxy, and during which the chapel's founder, archbishop don Pedro Tenorio (r.1377-1399), had himself been temporarily forced to flee from Castile.[4] The Creed often introduced medieval Castilian wills as a

---

All photos are reproduced with the kind permission of *Iberdrola* and *El Viso* publications.

[1] E. Juliá Martínez, "Una vida y una muerte", *Toletum* 57 (1939). Polo Benito was beatified by Pope Benedict XVI in October 2007.

[2] See J. Álvarez Lopera, "Realidad y Propaganda: el patrimonio artístico de Toledo durante la Guerra Civil", *Cuadernos de arte e iconografía* 3:6 (1990), 91-108; A. Beevor, *The Battle for Spain: the Spanish Civil War 1936-1939* (London, 2007), 57, 71, 107, 136-37, 209, 380-81, and 449.

[3] J. Polo Benito, *Pinturas murales de la Capilla de San Blas en la Catedral Primada de Toledo* (Toledo, 1925).

[4] See J. Valdeón Baruque, *Enrique II de Castilla: la guerra civil y la consolidación del régimen, 1366-1371* (Valladolid, 1966). Tenorio's life: E. Narbona, *Vida y*

formulaic statement of belief: thus began Tenorio's own will, or indeed that of king Enrique III of Castile, which was read aloud in the St. Blaise chapel in 1406.[5] Yet the unusual graphic realisation of the Creed in the St. Blaise chapel represents a particularly public articulation of Tenorio's orthodox faith, and in this essay I will investigate how art came to acquire a central role in such articulations in late medieval Castile.

The last three decades of the fourteenth century were a time of dramatic change for the kingdom of Castile-León, and particularly Toledo, the ancient city at Castile's geographic and political centre. Following a devastating civil war that had threatened the survival of Toledo's large and long-standing Jewish community, the victorious claimant of the bastard Trastámara line, Enrique II, had in 1374 founded a new royal chapel in the western bays of Toledo cathedral in a clear bid to associate his rule with the old *urbs regia*.[6] The heraldry of Enrique's wife, Queen Juana Manuel (d. 1381), in the keystones of the nave's western vaults, provides the clearest proof that the construction of Toledo cathedral, begun in the 1220s, reached completion in that period.[7] Completion of this long and expensive project allowed archbishop Tenorio to channel the income of the fabric into subsidiary projects, including the construction of a new cloister, library and choirscreen. Beyond the cathedral's walls, Tenorio also built a new bridge across the river Tagus for pilgrims to the Marian shrine at Guadalupe, rebuilt the great St. Martin bridge that gave access to Toledo from the west, and installed ascetic Hieronymites in the monasteries of St. Blaise at Villaviciosa and St. Catherine in Talavera. In this essay I will focus on Tenorio's funerary chapel of St. Blaise, introducing

---

hechos de D. Pedro Tenorio, Arzobispo de Toledo (Madrid, 1624) and L. Suárez Fernández, "D. Pedro Tenorio, Arzobispo de Toledo (1375 [*sic*]-1399)", in *Estudios dedicados a M. Pidal*, vol. 4 (Madrid, 1953), 602-627.

[5] Wills: Narbona, "Vida", f. 121r; C. Rosell, ed. *Crónicas de los reyes de Castilla desde don Alfonso el Sabio hasta los Católicos don Fernando y doña Isabel* (Madrid, 1877), 264-65.

[6] J. Valdeón Baruque, "La judería toledano en la guerra civil de Pedro I y Enrique II", in *Toledo Judaico. Papers presented at the 2nd Medieval Symposium. Centro Universitario de Toledo*, vol. 1 (Madrid, 1973), 107-131 and Rosell, "Crónicas", 39.

[7] The towers would not be complete for another hundred years, however; see D. Heim and A. Yuste Galán, "La torre de la catedral de Toledo y la dinastía de los Cueman. De Bruselas a Castilla", *Boletín del seminario de estudios de arte y arqueología* 64 (1998), 229-253. For a new chronology of construction for Toledo cathedral, see T. Nickson, "La Catedral: su Historia Constructiva", in *Libro de la Catedral Primada de Toledo*, ed. Ramón Gonzálvez Ruiz (Toledo, 2010), 148-161.

it first by way of a little-known project of 1387 that indicates the ambition of the artistic patronage of Tenorio, who was primate of Spain and advisor to several Popes, kings, and cardinals.[8]

In 1920, Francesc Almarche Vázquez published parts of a contract of 1387 between archbishop Tenorio and a Cypriot artist then resident in Valencia, "mestre Estevan Rovira de Xippre", for a vast new retable for Toledo cathedral.[9] Although recently published in its entirety, the contract has never been properly studied, doubtless because nothing seems to survive of the project, which may never even have been undertaken.[10] The ambition of the project is nonetheless noteworthy. Costing 4,500 gold florins, the retable's expense easily dwarfs that of the closest comparable projects, the 550 florins paid to Jacopo Starnina (1398) for a retable for the church of St. Augustine, Valencia, or the 650 florins paid to Juan Solano (1410) for an altarpiece in the Franciscan church in Zaragoza.[11] The retable was to stand behind the high altar and have a width of sixty-six *palmos*, "from pillar to pillar"; the height was to be twenty *palmos*, with the central panel higher by five *palmos* and with eight gables or pinnacles (*puntas*). Its broad, low format is similar to contemporary Castilian retables such as the Ayala altarpiece (by 1396) or that of Sancho de Rojas (1415)—both commissioned by Toledan patrons.[12]

---

[8] Tenorio's artistic patronage is analysed in chapter 5 of T. Nickson, *Toledo Cathedral: Art and Belief in Medieval Castile* (PhD thesis, Courtauld Institute of Art, 2009). The best published account is found in Á. Franco Mata, "El arzobispo Pedro Tenorio: vida y obra. Su capilla funeraria en el claustro de la catedral de Toledo", in *La Idea y el sentimiento de la muerte en la historia y en el arte de la Edad Media (II): ciclo de conferencias celebrado del 15 al 19 de Abril de 1991*, ed. Georges Duby (Santiago de Compostela, 1992), 73-94.

[9] F. Almarche Vázquez, "Mestre Esteve Rovira de Chipre. Pintor Trecentista Desconocido", *Archivo de arte valenciano* 6 (1920), 3-13.

[10] X. Company, J. Aliaga, L. Tolosa and M. Framis, *Documents de la pintura valenciana medieval i moderna I: 1238-1400* (Valencia, 2005), no. 511. The collected and re-painted fragments of a late fourteenth-century altarpiece now found elsewhere in the cathedral probably come from the Epiphany chapel at the west end of the nave. See Almarche Vázquez, "Mestre Esteve Rovira", 279, and the contract in the Archivo de la Catedral de Toledo, E.3.B.2.1, 5 November 1393.

[11] See J. Berg Sobré, *Behind the Altar Table. The Development of the Painted Retable in Spain, 1350-1500* (Columbia, 1989), 358-65.

[12] See Ibid., 27-48; 133-58; L. Lahoz, "De palacios y panteones. El conjunto de Quejana: imagen visual de los Ayala", in *Exposición Canciller Ayala*, ed. Félix López López de Ullíbarri (Vitoria-Gasteiz, 2007), 46-103, 61-73; M. Piquero López, *La Pintura Gotica Toledana Anterior a 1450*, vol. 1 (Toledo, 1984), 59-233.

The Spanish *palmo*—the length from the tip of the thumb to the tip of the little finger of an outstretched hand—is approximately 20.9 cm.[13] Sixty-six *palmos* would therefore equate to about 13.8 metres, equal to the distance between the Toledan presbytery piers that flanked the cathedral's high altar, making this the widest retable in medieval Iberia. The next largest, Jaume Huguet's retable for the church of St. Augustine, Barcelona (1460-86), measures sixty *palmos* across.[14] Unfortunately, the subjects of the retable's nineteen panels were not specified in the contract, but the project is nonetheless symptomatic of the new prominence accorded to panel paintings in late fourteenth-century Iberia, whilst its vast scale anticipates the great retables of late fifteenth-century Iberia. But before considering the motivations behind such an ambitious project, it will first be beneficial to return to the St. Blaise chapel.

In August 1389, archbishop Tenorio laid the foundation stone of a vast new cloister north of the cathedral, carved out of the clutter of subsidiary buildings and shops that had sprung up around the courtyard of the old mosque that had served as the city's cathedral from 1085 to the 1220s.[15] An ordinance of 1423 reveals that, as was typical in the peninsula, the cloister's primary function was funerary, and the cloister thus acts as a grandiose prelude to the private funerary chapel that Tenorio founded off its northeast corner in the same period.[16] Tenorio's effigial tomb here occupies the centre of a spacious cuboid chapel with an octagonal vault, an arrangement common in earlier episcopal claustral chapels at Burgos, Pamplona, and Salamanca cathedrals (see fig. 1).[17] The Toledan chapel's

---

[13] A. Sánchez Martín, "Pesos y medidas medievales y modernas en Medina del Campo y su tierra", in *Historia de Medina del Campo y su tierra*, ed. Eufemio Lorenzo, vol. 1 (Valladolid, 1986), 405-414; Berg Sobré, "Behind the Altar Table", 33.

[14] See A. Durán Sanpere, *Barcelona i la seva història*, vol. 3 (Barcelona, 1973-5), 161, n. 11; Berg Sobré, "Behind the Altar Table", App. 1.

[15] Narbona, "Vida", Book 2, i; P. León Tello, "El Alcaná de Toledo entre los bienes del arzobispo don Pedro Tenorio", *Revista de archivos, bibliotecas y museos* 75 (1968-72), 127-139.

[16] See M. J. Lop Otín, *El Cabildo Catedralicio de Toledo en el siglo XV. Aspectos institucionales y sociológicos* (Madrid, 2003), 111, n. 72; E. Carrero Santamaría, "El claustro funerario en el medievo o los requisitos de una arquitectura de uso ceremonial", *Liño: Revista anual de historia del arte* 12 (2006), 31-43.

[17] See M. Piquero López, *Las catedrales góticas castellanas* (Salamanca, 1992), 124; J. Martínez de Aguirre, *Arte y monarquía en Navarra, 1328-1425* (Pamplona, 1987), 261-63; and D. Sánchez y Sánchez, *La catedral vieja de Salamanca* (Salamanca, 1991), 139-46.

walls and octagonal vault are covered by bright narrative frescoes, as shown in fig. 2. The lower west wall shows the Last Judgement, whilst the remaining lower walls show scenes from the lives of saints Peter, Paul and Blaise.[18] A seventeenth-century biography of Tenorio also mentions scenes from the lives of saints Eugene and Anthony Abbot on two walls now substantially damaged by water seepage, together with an altarpiece showing Tenorio kneeling before Blaise, also now lost.[19] The upper walls show the New Testament scenes that Polo Benito connected to the articles of the Creed. Surviving account books for the chapel indicate that it was virtually complete by Tenorio's death in May 1399, although work continued there until 1406.[20]

Much relevant documentation on the chapel still survives and has been well studied. We know, for instance, that it was constructed under Rodrigo Alfonso, member of the important local family of stonemasons that had long worked on the cathedral.[21] The authorship of the frescoes is not so clear, however, especially given their numerous restorations and repaintings.[22] Yet notwithstanding a faded inscription recording that some work was done by "Rodriguez de Toledo, painter", it has never been doubted that the frescoes were directed by a Tuscan master, who may have come to Toledo via Valencia (Tenorio himself had spent considerable time in Perugia and Rome).[23] The identity of this master is problematic,

---

[18] The most important study of the frescoes is Franco Mata, "El arzobispo", op. cit.

[19] Narbona, "Vida", f. 103v.

[20] See A. Santa María and J. Carriazo, ed., *Cronica de Juan II de Castilla* (Madrid, 1982), 39; A. Sánchez-Palencia Mancebo, "La capilla del Arzobispo Tenorio", *Archivo Español de Arte* 48:189 (1975), 27-42, 29.

[21] G. Palomo Fernández, "Algunas precisiones y nuevos datos en torno a los Alfonso: una familia de canteros en el arzobispado de Toledo (1383-1431)", *Archivo Español de Arte* 73:292 (2000), 341-360.

[22] See A. Sánchez-Palencia Mancebo, "Pintores del siglo XV y Primera Mitad del siglo XVI en la Catedral Toledana. La Capilla de San Blas", *Anales Toledanos* 25 (1988), 57-80; idem, "La capilla de San Blas de la catedral de Toledo. Nuevos datos históricos", *Toletana* 14 (2006), 161-176; "La capilla de San Blas de la Catedral de Toledo", *Cuadernos de Restauración de Iberdrola* 11 (2005).

[23] F. Pereda and R. Gonzálvez, eds, *La Catedral de Toledo: descripción graphica y elegantissima de la S. Iglesia de Toledo, 1549, según el Dr. Blas Ortiz* (Toledo, 1999), 263-265. For relations between Toledo, Valencia and Italy, see E. Cruselles Gómez, "Mercaderes castellanos en Valencia (1400-1450)", in in *Actas de XV Congreso de la Corona de Aragón: Relaciones de la Corona de Aragón con los Estados cristianos peninsulares (siglos XIII-XV)*, vol. 2 (Zaragoza, 1997), 85-100; D. Igual Luis, "Los mercaderes italianos y las relaciones económicas entre

however: a document of 1393, in which the Florentine painter Gherardo Starnina was commissioned to paint a "panno" for the chapel of the Epiphany in Toledo cathedral (along with Nicolás de Antonio and Simón de Francisco), would provide a probable Tuscan candidate, except that the frescoes bear little resemblance to Starnina's reconstructed corpus.[24] Whatever his precise identity, a specifically Pisan origin for this master seems likely, however: the unusual depiction of the Last Judgement, in which the Virgin and Christ sit level in their own *mandorlas*, is similar to the mid-fourteenth-century Last Judgement in the Camposanto in Pisa, whilst the crowded figures of the Crucifixion scene (fig. 3) compare closely with a panel of the same subject by Antonio Veneziano (Starnina's master), now in the city museum in Pisa.[25]

No matter the painter's origin, it is clear that the fresco "programme" was closely supervised by the patron himself, as the selection of saints' lives is appropriate for a Toledan archbishop, and especially Tenorio. St.

Valencia y Castilla en el siglo XV", in *Actas de XV Congreso de la Corona de Aragón con los Estados cristianos peninsulares (siglos XIII-XV)*, vol. 2 (Valencia, 1997), 135-152.

[24] F. A. F. Vallejo, *Memorias y dissertaciones que podrán servir al que escriba la historia de la Iglesia de Toledo*, Madrid, Real Academia de Historia, Colección San Román, 2/Ms 24, 133; C. Torroja Menendez, "Pintores Florentinos en Toledo", *Historia y Vida* 79 (1974), 94-97, 95. On Starnina: Á. Vegue y Goldoni, "La dotación de Pedro Fernández de Toledo y Gerardo Starnina", *Archivo Español de Arte* 6 (1930), 277-279; idem, "Gerardo Starnina en Toledo", *Archivo Español de Arte* 6 (1930), 199-203; C. R. Post, *A History of Spanish Painting*, vol. 2 (Cambridge, MA, 1930), 222-32; J. van Waadenoijen, "A Proposal for Starnina: Exit the Maestro del Bambino Vispo", *Burlington Magazine* 116:851 (1974), 82-91; F. Sricchia Santoro, "Sul soggiorno spagnolo di G. Starnina e sull'identita del 'Maestro del Bambino Vispo'", *Prospettiva* 6 (1975), 11-29; A. T. Lurie, "In Search of a Valencian Madonna by Starnina", *Bulletin of the Cleveland Art Museum*, 76 (1976), 334-373; J. van Waadenoijen, "Ghiberti and the Origin of his International Style", in *Lorenzo Ghiberti Nel Suo Tempo: Convegno Internazionale di Studi*, vol. 1 (Florence, 1980), 81-87; G. Vasari, *Lives of the Painters, Sculptors and Architects*, vol. 1 (London, 1996), 224-27; F. Benito Doménech and J. Gómez Frechina, *La memoria recobrada: pintura valenciana recuperada de los siglos XIV-XVI* (Valencia, 2005), cat. 10; Piquero López, " La Pintura Gotica", vol. 1, 213. Starnina was also linked to Estevan of Cyprus via a Florentine merchant in Valencia: F. Bologna, "Un altro pannello del 'retablo' del Salvatore a Toledo: Antonio Veneziano o Gherardo Starnina", *Prospettiva* 2 (1975), 43-52.

[25] See D. Angulo Iñiguez, "La pintura trecentista en Toledo", *Archivo Español de Arte*, 7:19 (1931), 23-29, 24; Bologna, "Un altro pannello"; E. Carli, *Il museo di Pisa* (Pisa, 1974), no. 48.

Eugene was reputedly Toledo's first bishop and his feast was celebrated with special pomp in the cathedral; his cult had recently been boosted by the donation of new relics in 1348.[26] Tenorio's support of the eremitic Hieronymites probably explains his devotion to the hermit saint Anthony Abbot, who also appears on Tenorio's seal.[27] Likewise, Blaise (also a hermit saint) was a popular model of episcopal virtue in fourteenth-century Castile, and Toledo cathedral had several local hymns to him, together with a relic donated by Gil de Albornoz, archbishop of Toledo from 1338 to 1350.[28] Peter was Pedro's onomastic saint but also an emblem of his power: as for all medieval prelates, it was customary for Toledan archbishops to cite the authority of Peter and Paul when issuing indulgences, for instance.[29] It may be significant that Tenorio had been instrumental in supporting that apostolic authority, confirming his support of the Avignonese Pope Clement VII in May 1391.[30]

The clearest evidence of Tenorio's presence is, of course, his alabaster tomb at the chapel's centre. An inscription records that this was carved by Pedro Ferrandes de Toledo, "painter and sculptor [*entallador*]", who was also responsible for the tomb of Tenorio's nephew, Vicente Arias (d.1413) in the same chapel, and for a large group of similar tombs in the local

[26] See P. Pansier, "Histoire du monastère de Sainte Praxède d'Avignon (1346-1587)", *Annales d'Avignon et du Comtat Venaissin*, 4 (1916), 33-146, 81 and M. Revuelta Tubino, *Inventario artístico de Toledo. La catedral* (Madrid, 1989), 2 vols, vol. 1, 318. For St. Eugene's cult in the cathedral see Nickson, "Toledo Cathedral", op. cit., chapter 3.

[27] See R. Sánchez Sesa, "Don Pedro Tenorio y la reforma de las Órdenes monásticas en el último tercio del siglo XIV", *En la España medieval* 18 (1995), 289-302 & B. Fuentes Isla, "Las imagen de la Virgen en los sellos (estudio de la sigilografía española de los siglos XIII, XIV y XV)", *Revista de archivos, bibliotecas y museos* 3rd series, 27 (1923), 153-185 and 320-340, 163.

[28] G. M. Dreves, ed. *Hymnodia Hiberica* (New York and London, 1894), vol. 16, nos. 141-42; P. León Tello, *Judíos de Toledo*, vol. 2 (Madrid, 1979), no. 615.

[29] See, for example, the grant of pardon by Gil de Albornoz, 1 Sept 1345, citing the "poderío e abtoridat que *h*avemos de los bienaventurados sanct Pedro e sanct Pablo": Archivo de la Catedral de Toledo A.8.D.1.8, f. $5v_{13-16}$.

[30] Suárez Fernández, "D. Pedro Tenorio", 606. See M. Melero Moneo, "Iconografía de San Pedro y propaganda papal en la pintura gotica: Pedro y Simon el Mago como imagen de la confrontación del papa aviñones y el anti-papa romano. Un ejemplo catalano-aragones del siglo XIV", *Gesta* 36 (1997), 107-121.

area.[31] As is typical in the peninsula, both tombs bear repeated heraldry on their bases, here echoed by Tenorio's arms in the keystone of the chapel vault above.[32] Interspersed with rich foliate decoration, these heraldic displays act in some ways as surrogates for the dynastic continuity that Tenorio's office made impossible.

On Tenorio's tomb his arms alternate with busts of various saints, including St. Peter, further suggesting the kind of holy company that he kept. Around his effigy a Castilian inscription reads: "Here lies archbishop don Pedro Tenorio of praiseworthy memory, Archbishop of Toledo, Primate of the Spains, may God accept him into holy glory".[33] The recumbent effigy thus acts as both a prompt for intercessory prayers and a reminder of bodily resurrection at the Last Judgement, itself prominently depicted on the chapel's nearby western wall.

A Last Judgement scene similarly dominates the frescoed funerary chapel of bishop Rodrigo Díaz (d.1339) in the St. Martin chapel in Salamanca's old cathedral, and such eschatological concerns are typical of medieval tombs.[34] Yet the Toledan example is further glossed by two inscriptions that elicit Tenorio's particular concerns for reform.[35] One, in Latin, warns the seven chaplains endowed by Tenorio to be diligent in their duties: "By God's divine justice anyone who negligently talks during the divine office will die without a word."[36] Another inscription, this time

---

[31] S. R. Parro, *Toledo en la mano*, vol. 1 (Toledo, 1857), 690-91; T. Pérez Higuera, "Ferrand González y los Sepulcros del Taller Toledano (1385-1410)", *Boletín del seminario de estudios de arte y arqueología* 44 (1978), 129-142.

[32] F. Menéndez Pidal de Navascués, "La heráldica en las artes decorativas del Medievo", *Boletin de la real academia de buenas letras de Barcelona* 39 (1983-4), 63-82, 72, links this preference for heraldry to the use of textiles, and notes that the oldest Spanish tomb to bear extensive heraldry belongs to don Nuño in Las Huelgas, Burgos, c. 1209.

[33] Parro, "Toledo en la mano", vol. 1, 690.

[34] See Post, "A History of Spanish Painting", vol. 2, 142-47; D. Sánchez y Sánchez, *La catedral vieja de Salamanca* (Salamanca, 1991), 110-16; F. Gutiérrez Baños, *Aportación al estudio de la pintura de estilo gótico lineal en Castilla y León: precisiones cronológicas y corpus de pintura mural y sobre tabla*, vol. 2 (Madrid, 2005), 153-60. Jerrilynn Dodds has recently argued that the painted Last Judgement imagery in the Toledan church of San Román should be closely connected with the burial there of a local nobleman. See J. Dodds, "Rodrigo, Reconquest, and San Román: Some Preliminary Thoughts", in *Spanish Medieval Art: Recent Studies*, ed. Colum Hourihane (Princeton, NJ, 2007), 215-244.

[35] Reform: Suárez Fernández, "D. Pedro Tenorio", 605.

[36] Parro, "Toledo en la mano", vol. 1, 689.

in Castilian, addresses itself also to the many secular figures who sometimes gathered in this semi-private space:[37] "He who is guilty of pride, gluttony, lust, treachery, bad speech, whoring, stinking simony or wretched hypocrisy will join Satan, Lord of Hell, in fiery pain and sorrow of body and soul."[38] This colourful list recalls the litany of sins identified in the parody of the liturgy in the *Book of Good Love*, written by a local cleric, Juan Ruiz, in the 1340s, and testifies to the long-standing pastoral concerns of Toledan archbishops.[39]

If the saints in the chapel's lower levels may be understood as models for Tenorio's life and rule, those of the upper walls declare the orthodoxy of his faith. The scenes show the Annunciation, Nativity, Transfiguration, Christ before Pilate, Crucifixion, Entombment, Christ in Limbo, Resurrection, Ascension, Pentecost, God the Father and Christ enthroned, Christ judging, and the Resurrection of the Flesh. Polo Benito's suggestion that this represents more than a conventional depiction of Christ's life was prompted by the unusual image of Christ and God the Father enthroned (fig. 4), which has below it an excerpt from the Creed: "And He sitteth on the right hand of the Father" (Hebrews 1:3). Contemporary Creed cycles are rare, but this does seem the most plausible explanation for the rather unusual combination of scenes, even if one must therefore accept that the Transfiguration stands for "the life everlasting".[40] The Toledan context of

---

[37] See, eg, Rosell, "Crónicas", 262.
[38] Vegue y Goldoni, "Gherardo Starnina en Toledo", 5.
[39] See O. Green, "On Juan Ruiz's Parody of the Canonical Hours", *Hispanic Review* 24 (1958), 12-34; J. Ruiz and E. D. MacDonald (trans), *The Book of Good Love* (London, 1999), 99-103; J. L. Pérez López, "El 'Libro de buen amor' y la iglesia toledana", in *Actas del VIII Congreso Internacional de la Asociación Hispánica de Literatura Medieval*, eds. Margarita Freixas and Silvia Iriso, vol. 2 (Santander, 2000), 1443-1467, 1447.
[40] Post, "Spanish Painting", vol. 2, 224 and Franco Mata, "El arzobispo Pedro Tenorio", 88. Creed images: H. van Os and E. Biesta, *Vecchietta and the Sacristy of the Siena Hospital Church: A Study in Renaissance Religious Symbolism* (Gravenhage, 1974); F. Boespflug, "Autor de la traduction picturale du Credo à la fin du Moyen Age (XIIe-XVe siècle)", in *Rituels: mélanges offerts à Pierre-Marie Gy*, eds. Pierre-Marie Gyp, Paul de Clerck and Eric Palazzo (Paris, 1990), 55-84; D. Rigaux, "Un credo vénitien: les peintures murales de San Pietro di Feletto", in *Pensée, image et communication en Europe médiévale: à propos des stalles de Saint-Claude*, eds. Marie-Claude Mary and Eliane Vergnolle (Besançon, 1993), 119-128; Á. Franco Mata, "El Doble Credo en el arte medieval hispánico", *Boletín del Museo Arqueológico Nacional* 13:1-2 (1995), 119-136; J. Gregory, "The Credo

the programme is further emphasised by an abbreviated representation of the city in the scene of the Resurrection of the Flesh, in which the St. Martin bridge, rebuilt by Tenorio, is especially visible (fig. 5).[41] Thus we understand that by his good works and orthodox faith Tenorio was entitled to heavenly rest, and we can imagine a kind of narrative as the priest turned around at different moments of the mass. Looking west into the chapel he saw Tenorio's effigy and the Last Judgement fresco beyond it; turning east towards the altar he saw the altarpiece showing Tenorio kneeling before St. Blaise, himself the embodiment of the virtuous episcopal life.

The kinds of associations sketched above might be applied to many medieval funerary chapels, but the decoration of the St. Blaise chapel can also be understood according to the very particular historical circumstances of Castile in the late fourteenth century. Belief, namely orthodox belief, had long been a subject of concern to Toledo's archbishops, but such concern was especially intense after the dramatic events that took place in the summer of 1391, shortly before the chapel was painted.[42] In July 1391 the anti-Jewish pogroms that had begun in the Andalucian city of Seville reached Toledo with devastating effect. Toledo's large Jewish population had already suffered from attacks after the Black Death and during the Civil War in the 1350s and '60s, but the events of 1391 were to have repercussions that were felt in Spain even after all Jews were expelled in 1492. Some members of Toledo's large Jewish community fled, some were killed, some stayed true to their faith, but the vast majority were forced to undergo baptism and converted to Christianity—either in those violent and confusing events of 1391, or else in the following decades.[43]

---

of the Siena Cathedral Sacristy, 1411-12", *Renaissance Studies* 12 (1998), 206-227.

[41] Post, "Spanish Painting", vol. 2, 227.

[42] See D. Lomax, "El catecismo llamado de Albornoz", *Studia Albornotiana* 11 (1972), 213-232.

[43] For the period 1391-1412 see E. Mitre Fernández, *Los judíos de Castilla en tiempo de Enrique III. El pogrom de 1391* (Valladolid, 1994); P. López de Ayala and J.-L. Martín, ed., *Crónicas* (Barcelona, 1991); A. Pacios López, *La Disputa de Tortosa*, 2 vols. (Madrid, 1957); P. M. Cátedra, *Sermón, sociedad y literatura en la Edad Media: San Vicente Ferrer en Castilla (1411-1412). Estudio bibliográfico, literario y edición de los textos inéditos* (Valladolid, 1994); A. Echevarria, "Catalina of Lancaster, the Castilian Monarchy and Coexistence", in *Medieval Spain: Culture, Conflict, and Coexistence: Studies in Honour of Angus MacKay*, eds. Roger Collins and Anthony Goodman (Basingstoke, 2002), 79-122.

The scale of the conversions in Toledo is reflected by the massively reduced revenue received by the crown and Church from Toledo's Jewish community—revenue largely directed towards the endowments of the funerary chapels of Tenorio and Enrique II.[44] Such conversions had long been a dream of Spanish Christians, and seemed miraculous to many. In Valencia, so many Jews sought baptism that the clergy feared running out of chrism, until suddenly they found their vessels overflowing: "Consider for yourself", they asked the king, "whether these things can have a natural cause. We believe that they cannot, but can only be the work of the Almighty".[45]

These new *conversos* presented problems of their own, however, transgressing legal and social boundaries, and promoting growing anxiety about crypto-Judaism, the validity of conversions and the orthodoxy of *converso* faith. The Toledan anti-*converso* riots of 1449 and the establishment of the Inquisition in the following decades were direct consequences of these anxieties, and considerable criticism was later directed at the forced conversions of 1391.[46] In his widely disseminated *Fortress of Faith* (1458), for example, Alonso de Espina (who may himself have been a *converso*) emphasised the necessity of catechism before baptism, so that converts would truly understand Christian doctrine before their conversion.[47] Anticipating this spirit, in his *Memorial de Jesu Cristo*, which was written in the wake of the events of 1391, the little-known Toledan *converso* Juan el Viejo de Toledo stressed the importance of knowing the articles of the faith.[48] Meanwhile an anonymous *converso* scholar, long confused with Alfonso of Valladolid, used the Creed to begin his tract on the superiority of Christianity over other religions, structuring

---

[44] León Tello, "Los judíos", vol. 2, nos. 628; 636; 652.

[45] Cited in D. Nirenberg, "Enmity and Assimilation. Jews, Christians, and Converts in Medieval Spain", *Common Knowledge* 9 (2003), 137-155, 137.

[46] See E. Benito Ruano, *Los orígenes del problema converso* (Barcelona, 1976); S. Pastore, *Il Vangelo e la Spada. L'inquisizione di Castiglia e i suoi critici (1460-1598)* (Rome, 2003), 93.

[47] J.-L. Martín, "El sínodo diocesano de Cuéllar (1325)", in *Homenaje a Fray Justo Pérez de Urbel, OSB*, vol. 2, ed. Justo Pérez de Urbel (Santo Domingo de Silos, 1977), 145-176, 157; B. Netanyahu, *Toward the Inquisition: Essays on Jewish and Converso History in Late Medieval Spain* (Ithaca, N.Y., 1997), 43-75; A. Meyuhas Ginio, *De bello iudaeorum. Fray Alonso de Espina y su Fortalitium Fidei* (Salamanca, 1998), 95.

[48] *Memorial de Jesu Cristo* in Madrid, Biblioteca Nacional de España, Mss 9369, ff. 34r-38v.

his chapters according to the Creed's articles.[49] In this light, and despite its relative privacy, the Creed imagery in Tenorio's funerary chapel can be understood as commemorating Tenorio's orthodoxy whilst simultaneously celebrating his catechetical mission as pastor to a sudden new influx of *conversos* who were unschooled in Christian doctrine.

That the St. Blaise chapel frescoes specifically refer to these discourses of conversion and catechism is clear from their extensive depiction of the lives of St. Peter and St. Paul, notably Saul's conversion, Peter's refutation of Simon Magus, and Peter and Paul's preaching and baptism of new converts (fig. 6). And just as discussion of the status of converts inevitably looked back to Peter and Paul's debates on conversion in the New Testament, so too were Peter and Paul seen as models for the newly converted: the leader of Toledo's Jewish community, Rabbi Haym, took Peter's name, whilst Castile's most famous (and controversial) Jewish convert, Samel Ha-Levi, took Paul's.[50] What is particularly interesting, moreover, is that the frescoes include none of the overtly caricatured Jewish figures that one might expect. Their message of conversion and peaceful assimilation is, instead, typified by the prominent figure of Joseph of Arimathea at the Crucifixion, his status as a saintly convert emphasised by the polygonal halo that is commonly found on figures of Faith in Trecento painting (see fig. 3).[51]

There is, finally, one further way of understanding Tenorio's prolific artistic patronage. In the Toledan anti-*converso* riots of 1449, one of the three major complaints against the *conversos* was that they were idolatrous.[52] Questions of idolatry and the proper use of images had long vexed the Muslim, Jewish and Christian communities of medieval Europe and Castile in particular, and seem to have been especially heightened in the years around and after 1391.[53] During the great Christian-Jewish

[49] Pseudo-Alfonso de Valladolid, *Las tres creencias*, Madrid, Biblioteca Nacional de España, Mss 9302, f. 1v.

[50] León Tello, "Los judíos", vol. 1, 180; no. 38; vol. 2, nos 621; 636; L. Serrano, *Los conversos D. Pablo de Santa Maria y D. Alfonso de Cartagena: obispos de Burgos, gobernantes, diplomaticos y escritores* (Madrid, 1942).

[51] See M. J. Zucker, "The Polygonal Halo in Italian and Spanish Art", *Studies in Iconography* 4 (1978), 61-78; M. Levey, "Review: Count Seilern's Italian Pictures and Drawings", *The Burlington Magazine* 102 (1960), 122-123.

[52] Netanyahu, "Toward the Inquisition", 76-98.

[53] See C. Gilbert, "A Statement of the Aesthetic Attitude around 1230", *Hebrew University Studies in Literature and the Arts* 13 (1985), 125-152; M. Camille, *The Gothic Idol: Ideology and Image-Making in Medieval Art* (Cambridge, 1989), 129-

disputes at Tortosa in 1411/12, Christian scholars employed the old Gregorian argument to defend the didactic potency of images, recognising at the same time their power to *move*.[54] The preface to a famous Bible commissioned for a Toledan nobleman in 1422 contains an exchange of letters between a local rabbi and Toledan Franciscans that includes an extensive discussion of the propriety of Biblical illustration.[55] And Felipe Pereda has recently shown how the Castilian Church carefully promoted

---

94; G. Dahan, *Les intellectuels chrétiens et les juifs au Moyen Âge* (Paris, 1990), 500-3; J.-C. Schmitt, "La question des images dans les débats entre juifs et chrétiens au XIIe siècle", in *Spannungen und Widersprüche. Gedenkschrift für František Graus*, ed. Susanna Burghartz (Sigmaringen, 1992), 245-254; J. Maier, "Condiciones haláquicas previas en la confrontación religiosa del judaísmo con el cristianismo", in *Polémica judeo-cristiana: estudios*, eds. Carlos del Valle Rodríguez and Johann Maier (Madrid, 1992), 17-43, 27-30; G. Dahan, ed., *Inghetto Contardo. Diputatio contra Iudeos. Controverse avec les juifs* (Paris, 1993), 52-52; T. E. Burman, *Religious Polemic and the Intellectual History of the Mozarabs, c.1050-1200* (Leiden, 1994), 466-67; M. Lazar, "Anti-Jewish and Anti-*Converso* Propaganda: *Confutation libri talmud* and *Alboraique*", in *The Jews of Spain and the Expulsion of 1492*, eds. Stephen Haliczer and Moshe Lazar (Lancaster, CA, 1997), 153-236, 161; Meyuhas Ginio, "De bello iudaeorum", 44-45; R. Sánchez Ameijeiras, "'Ymagines sanctae': Fray Juan Gil de Zamora y la teoría de la imagen sagrada en las Cantigas de Santa María", in *Homenaje al Prof. García Oro*, ed. Miguel Romaní Martínez (Santiago de Compostela, 2002), 515-525; S. Moralejo, "D. Lucas de Tuy y la 'actitud estética' en el arte medieval", in *Patrimonio artístico de Galicia y otros estudios: homenaje al prof. Dr. Serafín Moralejo Álvarez*, vol. 2, eds. María Ángela Franco Mata and Serafín Moralejo (Santiago de Compostela, 2004), 299-302; S. Lipton, "'The Sweet Lean of His Head': Writing about Looking at the Crucifix in the High Middle Ages", *Speculum* 80 (2005), 1172-1207, 1179-81; A. García Avilés, "Imagenes 'vivientes': Idolatría y herejía en las Cantigas de Alfonso X el Sabio", *Goya* 321 (2007), 324-342; K. Kogman-Appel, "Christianity, Idolatry, and the Question of Jewish Figural Painting in the Middle Ages", *Speculum* 84 (2009), 73-107.

[54] A. Pacios López, *La Disputa de Tortosa*, vol. 2 (Madrid, 1957), 363. For Gregory's argument see M. Camille, "The Gregorian Definition Revisited: Writing and the Medieval Image", in *L'image. Fonctions et usages des images dans l'Occident médiéval* (Paris, 1996), 89-108; C. M. Chazelle, "Pictures, Books, and the Illiterate: Pope Gregory I's Letters to Serenus of Marseilles", *Word and Image* 6 (1990), 138-153.

[55] A. Paz y Melia, *La Biblia puesta en romance por Rabi Mosé Arragel de Guadalfajara, 1422-1433. La Biblia de la Casa de Alba* (Madrid, 1920), 2 vols, 15; S. Fellous, *Histoire de la bible de Moïse Arragel, Tolède 1422- 1433: quand un rabbin interprète la bible pour les chrétiens* (Paris, 2001).

the proliferation of religious imagery in the late fifteenth century in response to perceived *converso* critiques of images.[56] It is thus possible that archbishop Tenorio's conspicuous enthusiasm for religious imagery—evident in his funerary chapel, the vast new retable, and indeed other projects—should be seen as a defense of the proper use of images, a celebration of images according to the terms of Pope Gregory's famous defense of their use, and an articulation of the very substance of belief.[57]

---

[56] F. Pereda, "El debate sobre la imagen en la España del siglo XV: judios, cristianos y conversos", *Anuario del Departamento de Historia y Teoría del Arte, Universid Autonomía de Madrid* 14 (2002), 59-79; F. Pereda, *Las imágenes de la discordia: política y poética de la imagen sagrada en la España del cuatrocientos* (Madrid, 2007).

[57] These ideas are explored further in T. Nickson, *Toledo Cathedral: Art and Belief in Medieval Castile*, op. cit.

Illustrations

Figure 1, The St. Blaise chapel, Toledo cathedral.

Figure 2, Arrangement of scenes in the St. Blaise chapel.

North wall

<table>
<tr><td>Adoration of the Shepherds</td><td rowspan="2">Crucifixion</td><td>Entombment</td></tr>
<tr><td>Christ before Pilate</td><td>Christ in Limbo</td></tr>
<tr><td colspan="3">Lives of Eugene and Anthony Abbot (lost)</td></tr>
</table>

East wall

<table>
<tr><td>Resurrection</td><td>Matthew & Mark? (lost)</td><td>Ascension</td></tr>
<tr><td rowspan="2">Life of St. Blaise (lost)</td><td rowspan="2">Decapitation of St. Blaise</td><td>St. Blaise in jail</td></tr>
<tr><td>Burial of St. Blaise</td></tr>
</table>

South wall

<table>
<tr><td>God the Father & Christ enthroned</td><td rowspan="2">Pentecost</td><td rowspan="2">Resurrection of the Flesh</td></tr>
<tr><td>Christ Judging</td></tr>
<tr><td>Martyrdom of St. Peter? (lost)</td><td>Peter & Paul baptising; Peter enthroned</td><td>Conversion of St. Paul</td></tr>
<tr><td>Fall of Simon Magus</td><td>Door</td><td>Martyrdom of Peter and Paul</td></tr>
</table>

West wall

<table>
<tr><td>Transfiguration</td><td>Luke & John</td><td>Annunciation</td></tr>
<tr><td colspan="3">Last Judgement</td></tr>
</table>

Figure 3, St. Blaise chapel, west wall, detail: The Crucifixion.

Figure 4, St. Blaise chapel, south wall, detail: God the Father and Christ enthroned, Christ separating the blessed and the damned.

Figure 5, St. Blaise chapel, south wall, detail: The Resurrection of the Dead.

Figure 6, St. Blaise chapel, south wall. The bottom register shows Peter's martyrdom? (lost); Peter enthroned flanked by Peter and Paul baptising; the conversion of Saul.

Bibliography

"La capilla de San Blas de la Catedral de Toledo", *Cuadernos de Restauración de Iberdrola* XI (2005)

Almarche Vázquez, F. "Mestre Esteve Rovira de Chipre. Pintor Trecentista Desconocido", *Archivo de arte valenciano* 6 (1920), 3-13

Álvarez Lopera, J. "Realidad y Propaganda: el patrimonio artístico de Toledo durante la Guerra Civil", *Cuadernos de arte e iconografía* 3 (1990), 91-108

Angulo Iñiguez, D. "La pintura trecentista en Toledo", *Archivo Español de Arte* 7 (1931), 23-29

Beevor, A., *The Battle for Spain: the Spanish Civil War 1936-1939* (London, 2007)

Benito Doménech, F. and J. Gómez Frechina. *La memoria recobrada: pintura valenciana recuperada de los siglos XIV-XVI* (Valencia, 2005)

Benito Ruano, E. *Los orígenes del problema converso* (Barcelona, 1976)

Berg Sobré, J. *Behind the Altar Table. The Development of the Painted Retable in Spain, 1350-1500* (Columbia, MO, 1989)

Boespflug, F. "Autor de la traduction picturale du Credo à la fin du Moyen Age (XIIe-XVe siècle)", in *Rituels: mélanges offerts à Pierre-Marie Gy*, eds. Pierre-Marie Gyp, Paul de Clerck and Eric Palazzo (Paris, 1990), 55-84

Bologna, F. "Un altro pannello del 'retablo' del Salvatore a Toledo: Antonio Veneziano o Gherardo Starnina", *Prospettiva* 2 (1975), 43-52

Burman, T. E. *Religious Polemic and the Intellectual History of the Mozarabs, c.1050-1200* (Leiden, 1994)

Camille, M. *The Gothic Idol: Ideology and Image-making in Medieval Art* (Cambridge, 1989)

—. "The Gregorian Definition Revisited: Writing and the Medieval Image", in *L'image. Fonctions et usages des images dans l'Occident médiéval* (Paris, 1996), 89-108

Carli, E. *Il museo di Pisa* (Pisa, 1974)

Carrero Santamaría, E. "El claustro funerario en el medievo o los requisitos de una arquitectura de uso ceremonial", *Liño: Revista anual de historia del arte* 12 (2006), 31-43

Cátedra, P. M. *Sermón, sociedad y literatura en la Edad Media: San Vicente Ferrer en Castilla (1411-1412). Estudio bibliográfico, literario y edición de los textos inéditos* (Valladolid, 1994)

Chazelle, C. M. "Pictures, Books, and the Illiterate: Pope Gregory I's Letters to Serenus of Marseilles", *Word and Image* 6 (1990), 138-153

Company, X., J. Aliaga, L. Tolosa and M. Framis. *Documents de la pintura valenciana medieval i moderna I: 1238-1400* (Valencia, 2005)

Cruselles Gómez, E. "Mercaderes castellanos en Valencia (1400-1450)", in *Actas de XV Congreso de la Corona de Aragón: Relaciones de la Corona de Aragón con los Estados cristianos peninsulares (siglos XIII-XV)*, vol. 2 (Zaragoza, 1997), 85-100

Dahan, G., ed. *Inghetto Contardo. Diputatio contra Iudeos. Controverse avec les juifs* (Paris, 1993)

—. *Les intellectuels chrétiens et les juifs au Moyen Âge* (Paris, 1990)

Dodds, J. "Rodrigo, Reconquest, and San Román: Some Preliminary Thoughts", in *Spanish Medieval Art: Recent Studies*, ed. Colum Hourihane (Princeton, 2007), 215-244

Dreves, G. M., ed. *Hymnodia Hiberica* (New York and London, 1894)

Durán Sanpere, A. *Barcelona i la seva història*, vol.3 (Barcelona, 1973-5)

Echevarria, A. "Catalina of Lancaster, the Castilian Monarchy and Coexistence", in *Medieval Spain: Culture, Conflict, and Coexistence: Studies in Honour of Angus MacKay*, eds. Roger Collins and Anthony Goodman (Basingstoke, 2002), 79-122

Fellous, S. *Histoire de la bible de Moïse Arragel, Tolède 1422- 1433: quand un rabbin interprète la bible pour les chrétiens* (Paris, 2001)

Franco Mata, Á. "El arzobispo Pedro Tenorio: vida y obra. Su capilla funeraria en el claustro de la catedral de Toledo", in *La Idea y el sentimiento de la muerte en la historia y en el arte de la Edad Media (II): ciclo de conferencias celebrado del 15 al 19 de Abril de 1991*, ed. Georges Duby (Santiago de Compostela, 1992), 73-94

—. "El Doble Credo en el arte medieval hispánico", *Boletín del Museo Arqueológico Nacional* 13 (1995), 119-136

Fuentes Isla, B. "La imagen de la Virgen en los sellos (estudio de la sigilografía española de los siglos XIII, XIV y XV)", *Revista de archivos, bibliotecas y museos*, 3a época 27 (1923), 153-185 & 320-340

García Avilés, A. "Imagenes 'vivientes': Idolatría y herejía en las Cantigas de Alfonso X el Sabio", *Goya* 321 (2007), 324-342

Gilbert, C. "A Statement of the Aesthetic Attitude around 1230", *Hebrew University Studies in Literature and the Arts* 13 (1985), 125-152

Green, O. "On Juan Ruiz's Parody of the Canonical Hours", *Hispanic Review* 24 (1958), 12-34

Gregory, J. "The Credo of the Siena Cathedral Sacristy, 1411-12", *Renaissance Studies* 12 (1998), 206-227

Gutiérrez Baños, F. *Aportación al estudio de la pintura de estilo gótico lineal en Castilla y León: precisiones cronológicas y corpus de pintura mural y sobre tabla*, 2 vols. (Madrid, 2005)

Heim, D. and A. Yuste Galán. "La torre de la catedral de Toledo y la dinastía de los Cueman. De Bruselas a Castilla", *Boletín del seminario de estudios de arte y arqueología* 64 (1998), 229-253

Igual Luis, D. "Los mercaderes italianos y las relaciones económicas entre Valencia y Castilla en el siglo XV", in *Actas de XV Congreso de la Corona de Aragón con los Estados cristianos peninsulares (siglos XIII-XV)* (Valencia, 1997), 3 vols, vol. 2, 135-152

Juliá Martínez, E. "Una vida y una muerte", *Toletum* 57 (1939)

Kogman-Appel, K. "Christianity, Idolatry, and the Question of Jewish Figural Painting in the Middle Ages", *Speculum* 84 (2009), 73-107

Lahoz, L. "De palacios y panteones. El conjunto de Quejana: imagen visual de los Ayala", in *Exposición Canciller Ayala*, ed. Félix López López de Ullíbarri (Vitoria-Gasteiz, 2007), 46-103

Lazar, M. "Anti-Jewish and Anti-*Converso* Propaganda: *Confutation libri talmud* and *Alboraique*", in *The Jews of Spain and the Expulsion of 1492*, eds. Stephen Haliczer and Moshe Lazar (Lancaster, CA, 1997), 153-236

León Tello, P. "El Alcaná de Toledo entre los bienes del arzobispo don Pedro Tenorio", *Revista de archivos, bibliotecas y museos* 75 (1968-72), 127-139

—. *Judíos de Toledo*, 2 vols. (Madrid, 1979)

Levey, M. "Review: Count Seilern's Italian Pictures and Drawings", *The Burlington Magazine* 102 (1960), 122-123

Lipton, S. "'The Sweet Lean of His Head': Writing about Looking at the Crucifix in the High Middle Ages", *Speculum* 80 (2005), 1172-1207

Lomax, D. "El catecismo llamado de Albornoz", *Studia Albornotiana* 11 (1972), 213-232

Lop Otín, M. J. *El Cabildo Catedralicio de Toledo en el siglo XV. Aspectos institucionales y sociológicos* (Madrid, 2003)

López de Ayala, P. and J.-L. Martín, ed. *Crónicas* (Barcelona, 1991)

Lurie, A. T. "In Search of a Valencian Madonna by Starnina", *Bulletin of the Cleveland Art Museum*, 76 (1976), 334-373

Maier, J. "Condiciones haláquicas previas en la confrontación religiosa del judaísmo con el cristianismo", in *Polémica judeo-cristiana: estudios*, eds. Carlos del Valle Rodríguez and Johann Maier (Madrid, 1992), 17-43

Martín, J.-L. "El sínodo diocesano de Cuéllar (1325)", in *Homenaje a Fray Justo Pérez de Urbel, OSB*, ed. Justo Pérez de Urbel (Santo Domingo de Silos, 1977), vol. 2, 145-176

Martínez de Aguirre, J. *Arte y monarquía en Navarra, 1328-1425* (Pamplona, 1987)

Melero Moneo, M. "Iconografía de San Pedro y propaganda papal en la pintura gotica: Pedro y Simon el Mago como imagen de la confrontación del papa aviñones y el anti-papa romano. Un ejemplo catalano-aragones del siglo XIV", *Gesta* 36 (1997), 107-121

Menéndez Pidal de Navascués, F. "La heráldica en las artes decorativas del Medievo", *Boletin de la real academia de buenas letras de Barcelona* 39 (1983-4), 63-82

Meyuhas Ginio, A. *De bello iudaeorum. Fray Alonso de Espina y su Fortalitium Fidei* (Salamanca, 1998)

Mitre Fernández, E. *Los judíos de Castilla en tiempo de Enrique III. El pogrom de 1391* (Valladolid, 1994)

Moralejo, S. "D. Lucas de Tuy y la 'actitud estética' en el arte medieval", in *Patrimonio artístico de Galicia y otros estudios: homenaje al prof. Dr. Serafín Moralejo Álvarez*, vol. 2, eds. María Ángela Franco Mata and Serafín Moralejo (Santiago de Compostela, 2004), 299-302

Narbona, E. *Vida y hechos de D. Pedro Tenorio, Arzobispo de Toledo* (Madrid, 1624)

Netanyahu, B. *Toward the Inquisition: Essays on Jewish and Converso History in Late Medieval Spain* (Ithaca, N.Y., 1997)

Nickson, T. *Toledo Cathedral: Art and Belief in Medieval Castile* (Unpublished PhD thesis, Courtauld Institute of Art, 2009)

—. "La arquitectura catedralicia", in *Libro de la Catedral Primada de Toledo*, ed. Ramón Gonzálvez Ruiz (Toledo, 2010)

Nirenberg, D. "Enmity and Assimilation. Jews, Christians, and Converts in Medieval Spain", *Common Knowledge* 9 (2003), 137-155

Pacios López, A. *La Disputa de Tortosa*, 2 vols. (Madrid, 1957)

Palomo Fernández, G. "Algunas precisiones y nuevos datos en torno a los Alfonso: una familia de canteros en el arzobispado de Toledo (1383-1431)", *Archivo Español de Arte* 73 (2000), 341-360

Pansier, P. "Histoire du monastère de Sainte Praxède d'Avignon (1346-1587)", *Annales d'Avignon et du Comtat Venaissin* 4 (1916), 33-146

Parro, S. R. *Toledo en la mano*, 2 vols. (Toledo, 1857)

Pastore, S. *Il Vangelo e la Spada. L'inquisizione di Castiglia e i suoi critici (1460-1598)* (Rome, 2003)

Paz y Melia, A. *La Biblia puesta en romance por Rabi Mosé Arragel de Guadalfajara, 1422-1433. La Biblia de la Casa de Alba*, 2 vols. (Madrid, 1920)

Pereda, F. "El debate sobre la imagen en la España del siglo XV: judios, cristianos y conversos", *Anuario del Departamento de Historia y Teoría del Arte, Universid Autonomía de Madrid* 14 (2002), 59-79

—. *Las imágenes de la discordia: política y poética de la imagen sagrada en la España del cuatrocientos* (Madrid, 2007)

Pereda, F. and R. Gonzálvez, eds. *La Catedral de Toledo: descripción graphica y elegantissima de la S. Iglesia de Toledo, 1549, según el Dr. Blas Ortiz* (Toledo, 1999)

Pérez Higuera, T. "Ferrand González y los Sepulcros del Taller Toledano (1385-1410)", *Boletín del seminario de estudios de arte y arqueología* 44 (1978), 129-142

Pérez López, J. L. "El <<Libro de buen amor>> y la iglesia toledana", in *Actas del VIII Congreso Internacional de la Asociación Hispánica de Literatura Medieval*, vol.2, eds. Margarita Freixas and Silvia Iriso (Santander, 2000), 1443-1467

Piquero López, M. *La Pintura Gotica Toledana Anterior a 1450*, 2 vols. (Toledo, 1984)

—. *Las catedrales góticas castellanas* (Salamanca, 1992)

Polo Benito, J. *Pinturas murales de la Capilla de San Blas en la Catedral Primada de Toledo* (Toledo, 1925)

Post, C. R. *A History of Spanish Painting*, 6 vols. (Cambridge, MA, 1930)

Revuelta Tubino, M. *Inventario artístico de Toledo. La catedral*, 2 vols. (Madrid, 1989)

Rigaux, D. "Un credo vénitien: les peintures murales de San Pietro di Feletto", in *Pensée, image et communication en Europe médiévale: à propos des stalles de Saint-Claude*, eds. Marie-Claude Mary and Eliane Vergnolle (Besançon, 1993), 119-128

Rosell, C., ed. *Crónicas de los reyes de Castilla desde don Alfonso el Sabio hasta los Católicos don Fernando y doña Isabel* (Madrid, 1877)

Ruiz, J. and E. D. MacDonald, trans. *The Book of Good Love* (London, 1999)

Sánchez-Palencia Mancebo, A. "La capilla del Arzobispo Tenorio", *Archivo Español de Arte* 48 (1975), 27-42

—. "Pintores del siglo XV y Primera Mitad del siglo XVI en la Catedral Toledana. La Capilla de San Blas", *Anales Toledanos* 25 (1988), 57-80

—. "La capilla de San Blas de la catedral de Toledo. Nuevos datos históricos", *Toletana* 14 (2006), 161-176

Sánchez Ameijeiras, R. “'Ymagines sanctae': Fray Juan Gil de Zamora y la teoría de la imagen sagrada en las Cantigas de Santa María”, in *Homenaje al Prof. García Oro*, ed. Miguel Romaní Martínez (Santiago de Compostela, 2002), 515-525

Sánchez Martín, A. “Pesos y medidas medievales y modernas en Medina del Campo y su tierra”, in *Historia de Medina del Campo y su tierra* vol. 1, ed. Eufemio Lorenzo (Valladolid, 1986), 405-414

Sánchez Sesa, R. “Don Pedro Tenorio y la reforma de las Órdenes monásticas en el último tercio del siglo XIV”, *En la España medieval* 18 (1995), 289-302

Sánchez y Sánchez, D. *La catedral vieja de Salamanca* (Salamanca, 1991)

Santa Maria, A. G. d. and J. d. M. Carriazo, ed. *Cronica de Juan II de Castilla* (Madrid, 1982)

Schmitt, J.-C. “La question des images dans les débats entre juifs et chrétiens au XIIe siècle”, in *Spannungen und Widersprüche. Gedenkschrift für František Graus*, ed. Susanna Burghartz (Sigmaringen, 1992), 245-254

Serrano, L. *Los conversos D. Pablo de Santa Maria y D. Alfonso de Cartagena: obispos de Burgos, gobernantes, diplomaticos y escritores* (Madrid, 1942)

Sricchia Santoro, F. “Sul soggiorno spagnolo di G. Starnina e sull'identita del 'Maestro del Bambino Vispo'“, *Prospettiva* 6 (1975), 11-29

Suárez Fernández, L. “D. Pedro Tenorio, Arzobispo de Toledo (1375 [*sic*]-1399)”, in *Estudios dedicados a M. Pidal*, vol. 4 (Madrid, 1953), 602-627

Torroja Menendez, C. “Pintores Florentinos en Toledo”, *Historia y Vida* 79 (1974), 94-97

Valdeón Baruque, J. *Enrique II de Castilla: la guerra civil y la consolidación del régimen, 1366-1371* (Valladolid, 1966)

—. “La judería toledano en la guerra civil de Pedro I y Enrique II”, in *Toledo Judaico. Papers Presented at the 2nd Medieval Symposium. Centro Universitario de Toledo*, vol. 1 (Madrid, 1973), 107-131

Van Os, H. and E. Biesta. *Vecchietta and the Sacristy of the Siena Hospital Church: A Study in Renaissance Religous Symbolism* (Gravenhage, 1974)

Van Waadenoijen, J. “A proposal for Starnina: Exit the Maestro del Bambino Vispo”, *Burlington Magazine* 116 (1974), 82-91

—. “Ghiberti and the Origin of his International Style”, in *Lorenzo Ghiberti Nel Suo Tempo: Convegno Internazionale di Studi*, vol. 1 (Florence, 1980), 81-87

Vasari, G. *Lives of the Painters, Sculptors and Architects*, 2 vols. (London, 1996)
Vegue y Goldoni, Á. "La dotación de Pedro Fernández de Toledo y Gerardo Starnina", *Archivo Español de Arte* 6 (1930), 277-279
—. "Gerardo Starnina en Toledo", *Archivo Español de Arte* 6 (1930), 199-203
Zucker, M. J. "The Polygonal Halo in Italian and Spanish Art", *Studies in Iconography* 4 (1978), 61-78

CHAPTER SIX

POLYSEMY, METATHEATRICALITY, AND AFFECTIVE PIETY: A STUDY OF CONCEPTUAL BLENDING IN THE YORK PLAY *THE CRUCIFIXION*

KAREN WARD

The York play version of *The Crucifixion* is immensely metatheatrical since it repeatedly draws the audience's attention to its constructedness by breaking its own mimetic frame in various ways. My analysis focuses on the mimetic breaks caused by both the polysemous nature of the words "work" and "deed" in the play, and the mapping of the play and the cycle onto medieval spatio-temporal cultural realities that are recognizably consistent with their Biblical referents. The breaking of the mimetic frame, caused by either the polysemy of the play's words or by its contemporary mappings, would ideally result in a highly affective experience of the Passion and redemptive theology. Often, the breaking of the frame is metatheatrically humorous, which is problematically related to the grave subject matter of the crucifixion. This potentially blasphemous humour is actually what gives the play such affective potential. The instinctive response to humour is laughter, but laughter that is elicited from a representation of the crucifixion is likely to result in another emotional response in medieval viewers: guilt.

This humour-laughter-guilt progression serves the important medieval theological function of reifying the laity's understanding of their theoretical complicity in the crucifixion while masterfully avoiding the often didactic rhetorical modes of other literature written on the subject. In the York Cycle generally, and in *The Crucifixion* play specifically, the source domain (Biblical narrative) is effectively being mapped onto the target domain (medieval reality) to create a new blended cognitive space wherein the viewers are able to realize the applicability and relatedness of

Biblical events to their own subjective realities, regardless of their temporal estrangement from such events. The effect of the breaking of the represented mimetic frame (Biblical) in favour of the frame of reality (medieval life) helps the laity contextualize the Biblical events being represented theatrically and apply these narratives to elements of their own reality.

Conceptual blending is a theory that was developed by cognitive rhetoric scholars Mark Turner and Gilles Fauconnier. In a paper entitled "The Cognitive Study of Art, Language, and Literature", Turner describes conceptual blending as

> the mental operation of combining two mental packets of meaning—two schematic frames of knowledge or two scenarios, for example—selectively and under constraints to create a third mental packet of meaning that has new, emergent meaning.[1]

Through the various ways that the York Cycle and *The Crucifixion* break the mimetic frame of the Biblical stories they are portraying in mapping them onto medieval contemporary realities, many blends are created that serve several purposes, including to educate. Patricia Waugh, in her seminal and eponymous work on metafiction, defines the concept as "a term given to fictional writing which self-consciously and systematically draws attention to its status as an artefact in order to pose questions about the relationship between fiction and reality."[2] In this way, all metafiction (and metatheatre, in the case of the present analysis) can be seen as requiring the audience or readership to perform an act of conceptual blending. In metafiction and metatheatre, the concepts of fiction and reality are essentially being blended, and what emerges is a third mental space that is occupied by the significance of the relationship between these two seemingly opposing ideas.

Polysemy can be simply defined as the instance of a single word having several meanings or senses. There are many occurrences of meaningful polysemy within the York Cycle and in *The Crucifixion* that reflect the conceptual blending of Biblical drama and medieval realities. As I mentioned earlier, for this analysis I will focus on the polysemous nature of the Middle English words "werk" and "werkyng" as well as the word "dede" in the play. Towards the beginning of the play, Miles I says:

---

[1] Mark Turner, "The Cognitive Study of Art, Language, and Literature." *Poetics Today* 23 (2002), 9-20.

[2] Patricia Waugh, *Metafiction* (New York, 1984).

"Thanne to þis werke us muste take heede,/So þat oure wirkyng be noght wronge" (25-6).[3] The words "werke" and "wirkyng" are certainly polysemous in their uses here. According to the *Middle English Dictionary*, the word "werke" could simply mean "to act, behave or proceed", "to carry out one's assigned duties", "to carry out a command", or "to perform physical labour". It could also mean "to commit a wrongful or foolish act" or "to set events into motion". Each of these meanings has interesting and important relationships to the use of the word "werke" in *The Crucifixion.*

In one sense, the soldiers are simply carrying out the command given to them. In her essay "'The Education of a Torturer': The Psychological Impact of the York Crucifixion Pageant", Joan Faust convincingly relates Stanley Milgram's important 1963 psychological study, "The Perils of Obedience", to *The Crucifixion.* As Faust points out, the results of Milgram's study suggest that the average person is significantly more likely to inflict pain on another if ordered to by someone in authority. Faust argues that the play emphasizes the fact that the soldiers were ordered to complete this horrendous act, perhaps making the soldiers more forgivable from the audience's perspective.[4] Faust surprisingly does not cite the opening lines to the play, which read:

> Ȝee wootte yourselffe as wele as I
> How lordis and leders of owre lawe
> Has geven dome þat þis dote schall dye. (3-5)

This beginning immediately signals to the audience that the work being done by these soldiers has been ordered by "lordis and leders" and that the immorality of this work should not be projected onto the soldiers, who lack a certain amount of agency in relation to the task. The blame should be placed upon the officials. This distinction would be very cogent to a medieval audience, who while celebrating a religious holiday and having a day off from their daily work due to the feast of Corpus Christi, would perhaps be all the more aware of issues of work and authority. The audience is invited to experience suture with the soldiers, which is in essence a sort of conceptual blend. In their daily lives, the audience, like the soldiers, is given orders to complete certain tasks. By emphasizing the

---

[3] *The Crucifixion* in *Medieval Drama: An Anthology*, ed. Greg Walker (Oxford, 2000), 134-141. Subsequent references will refer to this edition.

[4] Joan Faust, "'The Education of a Torturer': The Psychological Impact of the York Crucifixion Pageant", *Journal of Popular Culture* 35 (2001), 155-168.

similarities between the requirements of labour of both the audience and the soldiers, playwright allows the audience to identify with the crucifiers of Christ. The conceptual blend involves both the audience and the soldiers, and results in a third mental space wherein the audience realizes their own destructive capabilities and inherent moral deficiencies from a medieval Christian perspective.

A Christian audience would have been well aware of the negative implications of the work of the crucifixion (i.e., committing a wrongful or foolish act), but they would have also been familiar with the fact that this work "set events into motion" for their own salvation. In the soldiers' work, then, the audience is presented with a paradox—the soldiers' work is evil, since they are crucifying the audience's saviour, but without the soldiers' work, there would be no salvation. By placing the work of the soldiers into this context, the playwright sanctions the audience's identification with the soldiers and their forming of a sort of suture, which, subsequently, will deeply affect the audience's theological sensibilities when they are presented with the visual representation of the soldiers' work when the cross is raised.

Some of the polysemy in the play is also quite humorous, but this humour, not unlike the suture between the audience and the soldiers acting out the crucifixion, can be theologically problematic. The line "Thanne to þis werke us muste take heede,/So þat oure wirkyng be noght wronge" (25-26) simultaneously implicates the work of both the soldiers being represented theatrically, and, in a metatextual breaking of the mimetic frame, the actors undertaking the performance. Thus, this line simultaneously means that the soldiers should pay attention to the work of crucifying Christ so that they do not make any technical errors, and that the actors staging the play should pay attention so that they will not miss any cues or forget their lines. While a medieval audience would possibly have laughed at the second sense of the word "werke," they would have realized the problematic moral and religious implications of their laughter due to the first sense of the word. This joke would have caused a sort of humour-laughter-guilt progression wherein the audience finds humour in a line, laughs, but then realizes the error of their response and experiences guilt.

This progression is ultimately caused by a conceptual blend. The coming together of the metatheatrically comedic and serious senses of the words "werke" would have essentially caused a highly complex and hybrid third mental space wherein the humour causes laughter, and the laughter causes guilt and subsequently the realization of the theoretical complicity of all of humanity in the crucifixion. By creating a blend of the serious and comedic senses of the word "werke," the playwright elicits a very complex and effective cognitive response from the audience that incorporates a wide range of emotional and rational functions. Because of the vast range of responses inspired by the blend, the polysemous use of the word "werke" in this way functions more effectively then directly and didactically telling the audience of their theoretical complicity in the act. Rather than making the audience understand this concept from a logical perspective alone, this blend allows the audience to experience their complicity emotionally as well.

There is another instance of the word "work" closer to the end of the play that also functions in a polysemously humorous way that is similarly problematic and probably would have caused the same humour-laughter-guilt progression. The task of crucifying Christ has proved to be extremely difficult for the soldiers throughout the play, as they find it both laborious and technically difficult. The soldiers complain of the amount of physical work that is required of them and of the errors and miscalculations that have been made with regards to the task. As Richard Homan points out:

> after one hand is nailed, [the soldiers] turn their attention to the other and find, 'It fails a foote and more! / The senuous are so gone inne' (lines 107-8). The first nail has pulled the sinews of Christ's hand into the nail-hole so far that his other hand now will not reach the other pre-drilled hole.[5]

Homan goes on to point out other errors in the soldiers' work, including the fact that they attach Jesus to the cross at the bottom of the hill and then carry him up, instead of having Jesus bear the cross up to the top of the hill and then attaching him to it there, as well as the fact that the soldiers make the mortise too big and have to correct the problem by using wedges to stabilize the cross. Homan states that the soldiers try to rationalize their errors by stating that, since these mistakes caused Christ more suffering, it was a better method of work.

Towards the end of the play, after the soldiers have completed the task of securing Christ to the cross, the first soldier asks Jesus: "Say, sir, how likes you nowe,/þis werke þat we have wrought?" (249-50). At first glance, Christ's reply seems to ignore the soldier's question:

> Al men þat walkis by waye or strete,
> Takes tente ȝe schalle no travayle tyne.

[5] Richard Homan, "Mixed Feelings About Violence in the Corpus Christi Plays", in *Violence in Drama*, ed. James Redmond (Cambridge, 1991), 93-100.

Byholdes Myn heede, Myn handis, and My feete,
And fully feele nowe, or ȝe fyne,
Yf any mournyng may be meete,
Or myscheve mesured unto Myne.
My Fadir, þat alle bales may bete,
Forgiffis þes men þat dois Me pyne.
What þei wirke, wotte þai noght;
Therfore, My Fadir, I crave,
Latte nevere þer synnys be sought,
But see þer saules to save. (253-264)

It seems that Christ's speech simply exonerates the soldiers from guilt, but does not speak to the question that the first soldier posed to Jesus before his speech. The co-operative principle put forward by Paul Grice states that subjects "make [their] contribution such as it is required, at the stage at which it occurs, by the accepted purpose or direction of the talk exchange in which [they] are engaged."[6] While there is a great deal of circumlocution demonstrated by the lengthy speech, Jesus certainly does adhere to this Gricean principle (though not to Grice's maxim of manner, which requires the speaker to be brief and avoid obscurity or ambiguity[7]), and responds to the soldier's question through another polysemous use of the word "wirke." In the line "What þei wirke, wotte þai noght", Jesus is simultaneously pleading with God to forgive the soldiers who crucify him, and making fun of their incompetence. In one sense, the line translates to the traditional understanding of Christ's words, which is "they know not what they do." But in a more direct translation, this line says "what they work, know they not," which implicates both a traditional, religious understanding of this line and a comedic understanding which suggests that Christ is complaining about the blatant incompetency of the soldiers who have crucified him.

Again, this is a joke at which a medieval audience is likely to have laughed, but this moment of humour carries with it even more serious problematic implications. This is the moment at which Christ's forgiveness and his suffering are being visually presented to the audience (note that at this point the cross, and thus the visual reminder of Christ's sacrifice, would have been raised and visible for the audience), and laughter at this point in the play, during the representation of divine altruistic love,

[6] Paul Grice, "Logic and Conversation", in *Syntax and Semantics*, vol. 3, eds. Peter Cole and Jerry L. Morgan (New York, 1975), 41–58.
[7] Raymond Gibbs, *The Poetics of Mind* (Cambridge, 1994).

suggests irreverence and impiety. Laughter at this point in the performance would surely have caused an even more profound experience of the humour-laughter-guilt progression mentioned earlier, and would have been highly effective in inspiring feelings of repentance amongst a medieval audience for their theoretical complicity in the crucifixion of Christ.

The word "dede" is thematically related to the word "werke" and also has meaningful polysemous uses in this play. The soldiers, as I mentioned earlier, often complain of the physical toil they experience while completing their task. Miles IV says: "þis dede for us will nevere be done" (196). Within this context, the word "dede" is polysemous in at least three ways. Firstly, "dede" can refer to the physical labour by the soldiers that is within the Biblical mimetic frame. Secondly, "dede" can refer to the repetitive task of the performance of the play by the actors; thirdly, in the most theologically charged level of meaning, "dede" can refer to the crucifixion of Christ, a task that, because of its gravity, will never be "done" for humanity. The first sense is seemingly the most literal interpretation of the line and is a mimetic representation of one of the soldiers who crucified Christ and who is complaining of the laboriousness of the deed. The soldier feels as though this work (the technical and physical work of crucifying Christ) is taking the group a very long time to complete, and he uses hyperbole to express his frustration. Certainly the actual labour of the crucifixion will eventually be finished, but the soldier uses exaggeration here for emphasis. In the second sense, the one in which "dede" refers to the acting of the play, the mimetic frame is broken and the actor complains of the repetitious act of staging the play multiple times.

According to a conjectural digital reproduction of the timing of the York Cycle with twelve stations, *The Crucifixion* pageant wagon would have began the Cycle at the Holy Trinity Priory at approximately 2:15pm and would have finished the day at the station known as Pavement at approximately 9:15pm.[8] This means that the actors would have been onstage for roughly seven hours. In addition to having a long day onstage and having to repeat the play between eight and seventeen times or more, the actors playing *The Crucifixion* had highly physical roles. For each performance, the actors would have to hoist the cross up (with the actor playing Jesus upon it), pretend to carry it up a hill, place the cross down when the play dictates that the soldiers become tired and rest on the hill, pick the cross back up while continuing to pretend that they are carrying it

[8] York Corpus Christi Pageant Simulator, "Simulator," http://jerz.setonhill.edu/resources/PSim/.

to the top of the hill, and then finally they must lower the cross into a mortise. They had to repeat these actions between eight and seventeen times or more depending on how many stations there were in the pageant that year.

Surely this would have been an exhausting task, as the cross used in the play would have been made of two very heavy beams of solid wood, and the actor upon it would have added weight. By a conservative estimate, the cross, with the actor on it, would have probably weighed more than 300lbs. Even with four actors lifting, this is very heavy, and the actors are required to carry the cross for extended periods of time, adding to their potential exhaustion. It is this second meaning that probably would have inspired laughter.

This metatheatrical joke of the actors complaining about the duration, repetition, and ardor that went into the production of this play was, if performed properly, likely to have caused laughter that was simultaneously religiously and socially problematic. The third meaning of this word suggests that the work of crucifying Christ will never be over for humanity, and the guilt of the sin will always be there. It is the third meaning of "dede" that makes laughter at the second meaning religiously problematic, and possibly blasphemous.

While the audience might laugh at the fact that the actors are having a hard time completing the Cycle, the third meaning of the word "dede" means that they are also laughing at humanity's responsibility for the crucifixion of Christ. The laughter would also have been somewhat socially problematic. Essentially, in laughing at the actors' plight, the audience is indulging in *schaudenfreude*. This seems dialectically opposed to the medieval Christina principal of *caritas*, or charitable love. Although a medieval audience would perhaps have instinctively laughed at this joke, the immediacy of guilt that would likely follow would certainly affect and edify the audience in a more meaningful way than would the same message communicated more didactically.

There is a great deal of humor, some of it also religiously problematic, implicitly built into the mapping of some of the Biblically mimetic details of the Cycle onto aspects of medieval culture with which these details are recognizably consistent. For instance, the guilds were each assigned to plays that were linked in some thematic way to their crafts or trades if possible. For example:

The Shipwrights were assigned to the play where God warns Noah to build an ark

The Mariners were assigned to the play where Noah, his family, and all the animals are sailing on the ark
The Chandlers (candle-makers) were assigned to the play where the shepherds find out about the coming of Jesus (the candles being symbolic for the North Star in this mapping)
The Vintners (wine-makers) were assigned to the play where Jesus turns water into wine
The Bakers were assigned to the play of the Last Supper
The Pinners (makers of hammers and nails) were assigned to the play of *The Crucifixion*

On this point, Anne Higgins rightly points out that

> in [...] these passion sequences grotesquely brutal scenes ingeniously exploit the figural possibilities of the guildsmen/actors' daily occupation with ropes and nails, mortices and wedges. The torturers of York [...] are characterized by enormous pride in their trade.[9]

The play of *The Crucifixion* has several metatheatrical moments of mimetic frame-breaking where the instruments of the crucifixion, items that the Pinners made, and which would have been used for other medieval work, are referred to and ostensibly "featured" in the play. Two such examples of this are: "And I have gone for gere goode speede,/Bothe hammeres and nayles, large and lange" (29-30), and: "Here are wegges arrayed / For þat, both grete and smale" (237-8). Higgins asks, "what can it mean to invent cruelties which pointedly use (or misuse) a producing craft's workaday skills, tools, or products?"[10] For the sponsoring guild, these lines essentially functioned as product placement for potential customers. Clearly this advertisement within the play is not without its complications; who would want their product associated with the suffering and/or crucifixion of Christ, especially in a social climate as religious as this one?

However, depending on the delivery of these lines, this "product placement" could be quite funny. The audience would see the nails, associate the production of the nails with the guild, possibly laugh due to the impropriety of the guild's hocking of the implements of the crucifixion, and then realize their error in laughing at this. It causes the

[9] Anne Higgins, "Work and Plays: Guild Casting in the Corpus Christi Drama." *Medieval and Renaissance Drama in England* 7 (1995), 76-97.
[10] Ibid.

same humour-laughter-guilt progression created by the polysemous use of the word “dede.” This realization would certainly prove more effective for edification than various other, more didactic rhetorical modes. The humour is produced by the incongruity of the conceptual blend of the medieval/commercial/guild meaning of the objects, and the devotional, religiously mimetic, and Christological significance of the nails and other tools as symbols of the torturing and crucifixion of Christ. In explaining the incongruity of some blends and the use of blends in irony, Mark Turner writes:

> We do not simply suppress lack of correspondence between source and target. On the contrary, information from the source that does not correspond to the target and cannot be projected onto the target is brought into the blended space exactly so we can understand the difference between the source and the target, and thereby recognize the clash. The projection is therefore significantly negative. We have seen that the projection is not direct; now we see that it is not essentially positive.[11]

It is a mixture of incongruity and inappropriateness that would have perhaps inspired laughter at this moment in the play, but it is the realization of the inappropriateness of this attitude towards the implements of the crucifixion that is the perhaps desired effect of hailing them in this overt way. The conceptual blend is also operating here in another, less humourous sense. Because this play was associated with the Pinners who made these nails, when the nails are visually presented to the audience as both artifacts of the work of their peers and as the nails used to crucify Christ in the Biblically mimetic representation they are watching, the audience is reminded that work is the punishment that was meted out to humanity by God as a result of original sin, and that all work, that of the Guilds’ and their own, is a result of this sin. The nails also represent God’s forgiveness and the salvation of humanity, since it is only through these nails that original sin can ever be forgiven. By blending the idea of medieval work and the work of crucifying Christ through the visual reminder of the nails, the playwright presents all of the complex Christian implications of work and salvation.

The plays also seems to have worked thematically with the geography of the city of York, as the last station in the cycle, known as Pavement, was the place in York where proclamations would have been read out and public executions conducted. Significantly, this is also the site where the

---

[11] Mark Turner, *The Literary Mind* (New York, 1996).

very last performance of the entire cycle, *The Last Judgement*, would have taken place. The theme of the play, judgement and justice, works with the extra-mimetic setting, which evokes justice to suggest the audience's implicitness in the on-stage action. This setting would have also worked well in this way for *The Crucifixion* once it reached Pavement. In this instance, the significance of place is entering into a conceptual blend with the significance of onstage action, and what emerges in the third mental space is the realization, for a medieval audience, of the relatedness of Biblical-historical events to their own material existence. The space, props, and actors of the plays are all being metatheatrically implicated in the Cycle's extra-mimetic hailing of them. By blending each of these elements of the Biblical content of the plays with their medieval context, what generally emerges is the medieval understanding not only of the stories themselves, but also of the ways in which the medieval laity were morally implicated in these stories.

Mark Turner writes that "blended spaces do cognitive work in the strongest sense. They provide inferences, emotions, and novel actions, and consequently leave their mark upon the real world."[12] The York Cycle, and *The Crucifixion* pageant within it, provide perfect examples of conceptual blending and the effect it can have upon real world, concrete realities. The conceptual blends required by the humour and the contemporary mappings of the plays produced very clear theological and ideological messages for their audiences, while not sacrificing their entertainment value. While the content of these plays would have been important in the religious education of the medieval laity, especially due to their production in the vernacular (when masses were held in Latin) and due to the inaccessibility to Bibles (which even when accessed would only be available in fragments, and not experienced chronologically), the *presentation* of the content was also extremely important in order to retain the audience's attention, especially since the entire cycle took about twenty hours to complete. By incorporating humour into the cycle, the playwrights effectively present their material in a way that would have been appealing to their audience. As Higgins concludes:

> The Corpus Christi drama represented figura in both a verbal and a visual demotic, by which even rude mechanicals could teach wide audiences how they all fit into the history of the world, and into the Christian economy of responsibility and grace. Their past was not dead, not their future unborn:

[12] Ibid, 74.

> both lived in the concrete reality of the guild productions of the Corpus Christi plays.[13]

By placing humour within the humour-laughter-guilt progression that I have discussed, the playwright integrates both edification and entertainment into the play and allows the audience to simultaneously enjoy the performance, learn about their faith, and understand their place within it.

Bibliography

Faust, Joan. "'The Education of a Torturer': The Psychological Impact of the York Crucifixion Pageant." *Journal of Popular Culture* 35 (2001), 155-168.

Gibbs, Raymond. *The Poetics of Mind* (Cambridge, 1994).

Grice, Paul. "Logic and Conversation," in *Syntax and Semantics*, vol. 3, eds. Peter Cole and Jerry L. Morgan (New York, 1975), 41–58.

Higgins, Anne. "Work and Plays: Guild Casting in the Corpus Christi Drama", *Medieval and Renaissance Drama in England* 7 (1995), 76-97.

Homan, Richard. "Mixed Feelings About Violence in the Corpus Christi Plays", in *Violence in Drama*, ed. James Redmond (Cambridge, 1991), 93-100.

The Crucifixion in *Medieval Drama: An Anthology*, ed. Greg Walker (Oxford, 2000), 134-141.

Turner, Mark. "The Cognitive Study of Art, Language, and Literature", *Poetics Today* 23 (2002), 9-20.

Turner, Mark. *The Literary Mind* (New York, 1996).

Waugh, Patricia. *Metafiction* (New York, 1984).

York Corpus Christi Pageant Simulator, "Simulator", http://jerz.setonhill.edu/resources/PSim/.

[13] Higgins, "Work and Plays", 92.

CHAPTER SEVEN

SCRIBAL EDITING IN *TUNDALE* AS THEOLOGICAL RHETORIC IN THE AGE OF ROMANCE

MICHAEL FOSTER

Scribal alterations to a source text, when they are not simply dismissed as errors, are often treated as blemishes to a presumed superior authorial text that has become corrupt through the human-mediated technology of manuscript transmission. Thus stemmatics traditionally privilege sources and usually serve to identify and eradicate scribal alterations to the text in order to get closer to the authorial voice.

As a result of this bias, the voices of the scribes who toiled to preserve and transmit the texts of the manuscript tradition that they produced are often eradicated or marginalized by modern editors in their tireless pursuit of authorial authenticity.[1] While such an editorial process is necessary to reproduce in print the presumed authorial intentions that made possible the text's genesis, it unfortunately loses important clues about the social history that the text reflected and produced as scribes made alterations they felt necessary or beneficial to an original source.

Most famously, Tim Machan has suggested a new approach to Middle English textual instability with his study of texts "as platforms to talk about larger issues of social practice".[2] Such an approach accepts the instability of Middle English texts and encourages a reading of scribal variants as conscious alterations to a malleable unit that could be manipulated and transformed to fit new contexts or new audiences. Such an approach uncovers the text's function as a representation of the

---

[1] This type of editorial bias is famously articulated by George Kane and E. T. Donaldson in their edition of *Piers Plowman, The B Version*, 2nd ed. (London, 1988), 128-220.

[2] Tim William Machan, *England in the Middle Ages* (Oxford, 2003), 19.

imaginative, interpretative, and contemplative efforts of scribes as they transmitted their texts to a new audience.

To demonstrate the fact that texts were edited by scribes over a historical process of emendation, revision, and censorship, I will examine the presentation of the Middle English *Tundale* with a particular emphasis on the ways in which scribes and compilers attempted to influence the manuscript audience's theological interpretation of the text. Such a reading suggests that scribes were confident enough with their ability to interpret and influence religious doctrine to emend a theological work for a new audience, although no evidence suggests that any of these scribes' labors were sanctioned by the Church.

It seems that medieval dogma as it was transmitted in the popular vernacular medium of Middle English was malleable enough to allow individual input from marginal voices far from centers of power or authority. It is thus in the corpus of vernacular religious texts that we can find spiritual diversity in the form of various scribe-editors' attempts to alter the religious experience of their secular readers.

Tundale: Sources and Analogues

Tundale is part of the medieval tradition of visionary literature that ultimately begins with the Book of Revelations. The fourth-century *St. Paul's Apocalypse* is an important contribution to the genre, although it has been considered apocryphal from St. Augustine onwards.[3] By the fifteenth century, various contributions to the tradition had made visionary literature a feature of the popular religious experience. These visions always promise a peek at otherworldly phenomena with varying degrees of sensationalism, authority, and spiritual depth.

Although the story of Tundale's vision was largely received in a vernacular adaptation amongst fifteenth-century English readers, it was originally produced for a more cloistered audience. The text began as a Latin account of an episode in the life of Tundale, a rich and wicked Irishman who briefly and partially dies at a dinner hosted by one of his debtors and is lead through various levels of post-mortem torment. This distinctively Irish vision of Hell began its textual life in 1149 or thereabouts, when a Benedictine monk from Ireland named Marcus arrived at the Cistercian St. James monastery in Regensburg, southern Germany,

[3] St. Augustine, *Tractates on John*, 98.8, edited in *Tractates on the Gospel of John 55-111*, ed. John W. Rettig (New York, 1994).

and wrote in Latin the *Visio Tnugdali.*[4]

Throughout the middle ages, Marcus' work or a derivative would be preserved in over 150 Latin manuscripts and translated or adapted into fifteen vernaculars, including Middle English. The story was read in every corner of Europe, from Iceland to the Balkans. Until post-medieval admiration of Dante's *La Divina Commedia*, no vision of the afterlife would travel so widely.

At its core and in its various versions, *Tundale* is a description of the afterlife from the perspective of Tundale. It is in many ways a "guided tour" in which the once-wicked protagonist—himself an everyman and an anti-hero, depicted as a flawed but ultimately redeemable human being—is taught that sins are eventually punished and that one cannot avoid the reckoning of God. Tundale is a loathsome medieval archetype—he is a moneylender, easily enraged, arbitrary and powerful, and contemptuous of the moral duties proscribed by the Church. At the same time, his horrific ordeal is often described with a sense of sympathy and his eventual reformation is a happy resolution that, by analogue, encourages readers to evaluate their own sins and to reform.

In addition to its obvious didacticism, the text provides a topographical description of Hell as a physical space populated with diverse sinners.[5] The horrors of the afterlife are described in vivid detail, and all Middle English versions tend to elaborate on the gruesome imagery of the Latin original. This evocative imagery can be considered the Middle English author's rhetorical adaptation of a more dour and contemplative original. For the Cistercians, Marcus produced what we might call a journalistic account of a metaphysical experience that he was familiar with from his native Ireland. Whether Marcus or his fellow monks in Germany believed that Tundale really existed and was privy to a sneak preview of the afterlife is obviously unknowable, but we have no reason to doubt the sincerity of their faith and Marcus' intention to make his fellow monks aware of a reported near-death experience and mystical vision.

Likewise, many fourteenth and fifteenth-century readers of the Middle English text would probably have taken this text seriously, since mystical visions had become a prominent feature of religious life by the end of the middle ages. However, the authority of the text in its Middle English

---

[4] Regensburg and Ireland were closely linked from the twelfth-century onwards. See Peter Harbison "Twelfth and Thirteenth Century Irish Stonemasons in Regensburg (Bavaria) and the End of the 'School of the West' in Connacht", *Studies: An Irish Quarterly Review* 64 (1975), 333-346.

[5] The history of purgatory's physical presence is famously discussed in Jacques Le Goff, *The Birth of Purgatory* trans. Arnold Goldhammer (Chicago, 1986).

context would have been markedly different from the Latin original, which was presented to a pre-existing community of monks at a relatively important centre of international learning by a newcomer from a peripheral and foreign (albeit well-established) Christian culture. The Middle English text is presented in its manuscript copies without explanation or introduction as a narrative account of a past event, and would probably have been treated as a somewhat authoritative document, although the fact that it was vernacular and presented in a text not distributed or produced by the Church may have made it somewhat suspect.

While the Latin original serves as a direct antecedent, there are numerous analogous texts from the late middle ages. The most obvious would be Dante's *Divine Comedy*, particularly the *Inferno* and *Purgatorio* sections, which describe a similar journey through various levels of posthumous suffering and punishment. Dante was aware of the visionary literature that had developed by his lifetime,[6] although the very popularity of the ideas, images, and themes of the visionary tradition makes it impossible to connect Dante to any specific text.[7]

Even if they are not immediately related, *Tundale* and Dante's masterpiece share enough general features to warrant a cursory comparison. Most immediately similar is the topographical nature of the afterlife; Hell, Purgatory, and Heaven are treated as real places that have a physical presence in relation to the material world. Also, in both the severity of each person's sins is given consideration and the relationship between the form and severity of punishment is put within the context of the sinner's indulgences. Presumably for the reader's sake, Tundale often asks his guide why the sinners deserve the type of punishment they receive. For example, upon seeing a sinner forced to traverse a bridge in Hell:

> Tundale askede þe angelle bryȝte:
> "What menes þys hydwys syȝte?"
> The angell onswered þus aȝeyn,
> "For þe & oþer ys þys payne
> That robbon men of her rychesse,
> Or takes ony good þat herys es,

[6] A very old but still useful study of the visionary tradition before Dante can be found in Alessandro D'Ancona, *I precursori di Dante* (Florence, 1874). See also Jan Emerson, *Imagining Heaven in the Middle Ages: A Book of Essays* (London, 2000) and Eileen Gardner, *Visions of Heaven and Hell Before Dante* (New York, 2008).

[7] However, C. S. Boswell asserts that *Tundale* did indeed influence Dante—see *An Irish Precursor of Dante: A Study of the Vision of Heaven and Hell Ascribed to the Eighth Century Irish Saint Adamnán* (London, 1908), 214-220.

Lewed or lerede or of holy Kyrke,
Or ony wronge to hym wryke.
Some han more payn & some lesse,
After þat her synne es (631-640)[8]

A sense of divine judgment is thus invisible behind the hideousness of Hell, and punishments are meted out with a sense of righting previous wrongs to an appropriate degree. This evaluative approach to a hierarchy of disparate sins demands an organization of Hell into levels, which Marcus creates with a topographical narrative structure. Although less organized and geometric, *Tundale's* physical topography anticipates the multi-dimensional complexity of Dante's later text, which also employs a similar interrogative analysis and topographical manifestation of the relative evils of sins. Sinners, thus, are simultaneously separated and unified in their punishments.

This structural organization is well articulated by Alison Morgan, who suggests that the "explicit separation of one class of sinner from another; the gradual increase in gravity of sin and corresponding torment as we travel deeper into the pit of hell" is reason to compare *Tundale* with Dante's vision, although such an Aristotlean categorization of spiritual phenomena is more likely the result of a "marriage between a large mass of traditional material and the Aristotelian categories" than it is proof that Dante was influenced by the Irish vision.[9] It is more likely that Marcus' and Dante's imaginations reflect a theoretical model of stratification that became the backbone of every aspect of medieval society and, as it manifested itself in textual narratives of the afterlife, this hierarchical model of classes of sinners and sins became part of the popular understanding of the afterlife as it affected real people.[10] The experiences of hierarchy in everyday life and the traditional doctrine against Christian sin complemented one another and, over time, began to transform the ways in which Hell and divine punishment were perceived.

To find a vision of the afterlife comparable to *Tundale*, we need not go so far afield as Italy: Ireland produced another visionary text in which a

---

[8] Edited in *The Vision of Tundale Edited from BL MS Cotton Caligula A.II*, ed. Rodney Mearns (Heidelberg, 1985).

[9] *Dante and the Medieval Other World* (Cambridge, 1990), 110-112. See also Teodolinda Barolini's discussion in *Dante and the Origins of Italian Literary Culture* (New York, 2006), 109-113.

[10] For a study of the visionary tradition as a form of theological narration, see Carol Zaleski, *Otherworld Journeys: Accounts of Near-Death Experience in Medieval and Modern Times* (Oxford, 1988).

journey to the next life presents a stern warning to readers. This story became popular through a retelling known as the *Tractatus de Purgatorio Sancti Patricii*, also written in Latin outside of Ireland in the twelfth century (ca.1180). Purgatory in this text is treated even more as a physical space than in *Tundale*, since it is directly connected to the Earth by a portal in a cave on Station Island in Lough Derg, County Donegal, Ireland. The author has been identified as Henry of Sawtry, a monk in Huntingdonshire, although virtually nothing else is known about him.[11] This text was translated independently into two Middle English versions, which modern editors and some manuscript compilers give the title "Owayne Miles" after the protagonist who is the focus of the Middle English text.

Both versions of *Oweyn Miles* are most immediately identifiable with the visionary tradition, but they also share the focus on a single noble knight, the vivid description of wondrous adventures, and the struggle of the knight against evil enemies that is reminiscent of the Middle English romances, which often accompany this text in manuscripts.[12] It is unsurprising that Marie de France found it fitting to be adapted as one of her *Breton lais*. Her rendition, *L'Espuragoire Seing Patriz*, is arguably the most enduring version of the story, and influenced Robert Southey's retelling.

Medieval versions of this story vary in theological emphasis. For example, the early fifteenth-century *Vision of William of Stranton* is, in the words of Takami Matsuda, "a markedly homiletic version with a detailed description of the infernalized torments", but Marie de France's *Espurgatoire Seint Patriz* treats the story as "a knightly adventure".[13] The Middle English *Owayne Miles* is somewhere between these two extremes; Owein's knightly virtues and the horrors of his experience combine to inspire awe in the afterlife and admiration for his bravery. His ordeal into the afterlife becomes an example for medieval readers on how to remain steadfast and brave in the shadow of death and the torments of Hell. Just as *Ypotis* (ff.79v-83) teaches wisdom, *Owayne* teaches bravery, and reminds readers that Purgatory, Heaven, and Hell are real places, as much a part of the landscape as any part of Ireland or the world. Such a medieval

---

[11] See Robert Easting, "The Date and Dedication of the Tractatus de Purgatorio Sancti Patricii", *Speculum* 53 (1978), 778-83 and Easting, ed., *St. Patrick's Purgatory* (Oxford, 1991), xliv.

[12] Particularly the Auchinleck manuscript (NLS Advocates 19.2.1) and BL Cotton Caligula A.ii , discussed below.

[13] Takami Matsuda, "*Sir Gawain and the Green Knight* and St. Patrick's Purgatory", *English Studies* 88 (2007) 497-505, at 504.

view of the world is well attested in the famous *mappa mundi* preserved in Hereford Cathedral.[14]

Owayne is thus an example of heroism against a real and dangerous evil, a common theme of the manuscript in religious and secular texts alike, and a reminder that the options of eternal torment and eternal bliss await the living. The author of the text makes this theme explicit almost immediately, when describing Patrick's preaching in Ireland:

Fyrste he precehd of heuen-blysse,
Whoeuur go þydur may ryght nowȝt mysse;
Sethen he preched of helle-pyne,
Howe we þem ys þat comeþ þerinne;
And þen he preched of purgatory,
As he fonde in hys story.
But ȝet þe folke of þys contre
Beleued not þat hyt myȝth be,
And seyde, but ȝyf hyt were so,
That eny mon myth hymself go
And se all þat and come ageyn,
Then wolde þey beleue fayn. (33-44)

The account of Owayne's descent and the existence of the entry to Purgatory serve as definite evidence, if the authority of the text is accepted, that the afterlife is a real, physical threat. The description of Owayne's journey, and the terror he feels as he is tormented by demons in Purgatory, bolsters fear in a place that is presented as very real and horrific. Owayne learns from his vision and ends by becoming a crusader to the "Holy Londe" (672) and, after dying as an old man, he reaches "þe blysse þat lastes aye" (679).

Significantly, *Owayne Miles* tells the story of a heroic knight, while *Tundale* is about a sinner who must be scared into reforming his evil habits. As a result, *Tundale* lacks the fantastical energy of the pseudo-romantic elements of *Owayne Miles* while possessing a more immediate rhetorical urge by reminding readers not only about the afterlife, but that real people just like them can and have faced horrendous torment as a result of their sins. *Tundale* thus appeals to a different part of the consciousness: for the believer, it is more like a hellfire sermon than it is a literary work for the imagination and the intellect. In this sense, it diverges from Dante's *Divine Comedy*, with its complex scientific, numerological, political, and ideological undercurrents.

While modern readers would most immediately think of Dante when

[14] P. D. A. Harvey, *Mappa Mundi: The Hereford World Map* (London, 1996).

approaching *Tundale*, it is unlikely that fifteenth-century English readers would have felt the same way. Dante's influence on English literature outside of the Chaucerian tradition would remain minor until early-modern interest in Italian culture inspired a return to the *Divine Comedy* and, of course, no Middle English translation of the Italian text was ever produced. Instead, English medieval readers would have associated *Tundale* with two more immediate traditions: the visionary literature and the medieval hellfire sermon with its Church-sanctioned threats of punishment in the next world for sins in this one. Neither of these traditions is solely identifiable as a "literary" phenomenon in the modern sense, partly because they served as rhetorical methods of social guidance and control. While Marcus may not have intended his text to be a warning to the secular Christian community that their sins will be accounted for eventually, this is most likely how the Middle English translation was to be received centuries later.

Tundale: Surviving Manuscripts and their context

Unlike its Latin original, the Middle English *Tundale* seems to have appealed to secular readers who also had a taste for narratives about knights. It survives in five manuscripts:

1. London, British Library MS Cotton Caligula A. ii, ff. 95v-107v
2. London, British Library MS Royal 17.B.xliii, ff. 150-184v
3. Edinburgh, National Library of Scotland MS Advocates' 19.3.1, ff. 98-157v
4. Oxford, Bodleian Library MS Ashmole 1491 (*SC* 7656), a fragment of five ff. at the end of the book
5. Tokyo, Takamiya 32, ff.166v-175v

Of these, four also preserve romances and are surprisingly similar in content, character, period of production, and general design. I shall give a brief description of each manuscript which will form the basis for a codicological comparison that suggests a specific type of implied audience for the Middle English text and a context of textual reception markedly different from that of the Latin original.

For my study, I have chosen to focus on the text in Cotton Caligula A.II, produced in southern England sometime around 1460.[15] This manuscript contains a variety of religious texts, but it is most famous for

[15] This manuscript is a composite of two independent manuscripts that were bound together in the Cotton library before 1654 and given their present shelfmark. Part I (ff.3-139, olim Vespasian D.VIII) is my present concern.

its romances, including unique copies of *Sir Launfal* and *Emaré*, both of which seem considerably earlier than Caligula.[16] The presence of alliterative, tail-rhyme, ballad, and couplet stanzas demonstrate that the compiler may have wished to include a variety of poetry for the book's intended audience. In any case, metre was probably not as important to the compiler as the themes and subject matter of the texts.

The diversity of textual provenances in the book also suggests that the scribe may have intended to preserve a survey of provincial English texts, or he was unconcerned with (or ignorant of) the provenances of his texts, being more focused on subject matter. Northern texts in Caligula invariably travelled south for some time before the compiler copied them from his exemplar; this can be demonstrated particularly in *The Siege of Jerusalem*, the *Epistle of Susan*, *Chevalere Assigne*, and *Emaré*, which are textually much removed from more northern copies (or implied exemplars), and which have been translated to a more southern dialect.[17] These texts have the mark of many dialects from the south and Midlands, suggesting that they travelled widely before reaching Caligula.

The manuscript has no marks of ownership except some medical recipes scribbled on f.13v, which is too late (sixteenth century) to help us date the manuscript's production. The main scribe's Anglican and secretary letter-forms are standard for the mid-fifteenth century, but reveal little more. The earliest watermark was in use in about 1423 and the latest in about 1476, and the latest datable poem in the book is Lydgate's "The Nightingale" (ff.59-64), which was most likely written in 1446.[18] "O Mors Quam Amara Est Memoria Tua" (ff.57v-58) has been used to date Caligula, since it is also preserved in Harley 116, where we find an added stanza that refers to the death of Ralph, Lord Cromwell (d.1456).[19] However, Rosemary Woolf argues that, since this stanza disrupts the sense

[16] See A. J. Bliss, ed. *Sir Launfal* (London, 1960) and Edith Rickert, *Emaré* (London, 1908), respcctively.

[17] The most comprehensive study of this movement is in Ralph Hanna and David Lawton, eds., *The Siege of Jerusalem* (Oxford, 2003) xxv-lxviii.

[18] For this dating, see O. Glauning, ed., *Lydgate's Minor Poems* (Oxford, 1900), xxxvi-xxxviii and H. N. MacCracken, ed., *The Minor Poems of John Lydgate* vol. 1 (Oxford, 1911), xxxiii-xxxiv, who questions Lydgate's authorship.

[19] Frances McSparran, following Carleton Brown (*Religious Lyrics of the Fifteenth Century* (Oxford, 1939), 339-40), suggests that the Caligula text is a derivative of Harley—see *Octovian Imperator* (Heidelberg, 1979), 13. McSparran erroneously reports that Cromwell died in 1454.

of the poem, it is probably a later insertion of an independent textual history.[20] If this is the case, the poem cannot help us to date Caligula.

The Latin chronicle on ff.109-110v provides its own *terminus ad quem*. The manuscript's scribe has written only the first section, which ends before the reign of Edward IV, suggesting that the book was completed in 1460/1. Two secondary hands extend the chronicle to Richard III, suggesting that the manuscript was in continuous use until 1483-5. From this limited evidence, the manuscript can be dated to 1446-1461.[21] I would suggest that the manuscript was compiled closer to 1460/1, since the scribe could not have copied all of the chronicle any earlier, and the manuscript seems to be one continuous production written according to a plan over a relatively short period of time.

Establishing the manuscript's provenance is an equally challenging linguistic puzzle that has yet to be solved. The Latin chronicle and possibly the Latin tags in "Veni, Coronaberis" (ff.107v-108) suggest that Caligula's intended readership had some knowledge of Latin. The rest of the texts are in Middle English, and display heavy dialectical variation. Previous studies have argued that the scribe's dialect is Kentish or East Anglian, but, as M. L. Samuels observes,[22] this appears to be one stratum of many. Following this, I would suggest that the scribe employed a *Mischsprache*, and his dialect cannot be distinguished from the many dialects of his exemplars, many of which he may have transcribed with little dialectical alteration.

The Kentish theory is based on Mills's study of three poems that he ascribes to Thomas Chestre[23] on the basis of a shared and idiosyncratic "fondness for certain words, phrases, or formulae, to which they bear witness ... [and] this common material is used in a manner that betrays a consistent desire to meet the needs of rhyme, at whatever cost to stylishness and sense".[24] Mills attempts to determine the archetype of

[20] See Rosemary Woolf, *The English Religious Lyric in the Middle Ages* (Oxford, 1968), 340 and Easting, *St. Patrick's Purgatory*, xxvi. See also Thorlac Turville-Petre, "The Persecution of Elizabeth Swillington by Ralph, Lord Cromwell", *Nottingham Medieval Studies* 42 (1998), 174-87.

[21] Edith Rickert suggests 1446-1460 on this same basis (see *Emaré*, ix).

[22] "Kent and the Low Countries: Some Linguistic Evidence", in *Edinburgh Studies in English and Scots*, ed. A. J. Aitken, Angus McIntosh and Hermann Palsson (London, 1971), 3-26, at 13.

[23] M. Mills, "The Composition and Style of the "Southern" *Octavian*, *Sir Launfal*, and *Libeaus Desconus*', *Medium Aevum* 31 (1962), 88-109.

[24] M. Mills, ed., *Lybeaus Desconus* (London. 1969), 65. Bliss, *Sir Launfal*, 12-13 suggests the same. See the response by A. S. G. Edwards, "Middle English

Libeaus Disconus (ff.42v-57) and its author's language, which he identifies as Kentish. He also suggests that Caligula's treatment of the text seems to demonstrate the scribe's "own conservativism and the fact that an unbroken tradition of texts set down in very much the same region must link C with Chestre's original".[25]

In his edition of *Sir Launfal*, A. J. Bliss observes the development of OE æ/e, OE y, ȳ, and ǣ, the effacement of OE g before d, ǭu>au, and the use of *hoth* "heath" in line 250, often found in Kent and East Sussex place names, although this is in rhyme position and may be the poet's and not the scribe's dialect.[26]

Richard Jordan writes that the three Chestre-texts contain "very confused writing", and continues to suggest that the

> interchange of d and þ in MS Cotton Caligula A 2 can be explained on the assumption of the following geographically limited sound changes: the voicing of every initial /Θ/ to /ð/; the voicing of final /Θ/ in an unstressed syllable; and the change of every /ð/ in whatever position to /d/.

As a result, /ð/ would be completely eliminated, and /Θ/ would occur in only the three situations:

1. Finally in stressed syllables
2. Medially when geminated
3. Initially in words borrowed "too late to be affected by the change of initial /Θ/ to /ð/"[27]

These shifts may be the result of a flattening of various aspirations of /d/, /ð/, and /Θ/ due to dialectical contact, or, as has been suggested, they might reflect changes in orthographical conventions as well as linguistic change in fifteenth-century England.[28] The first and third shifts are common in place-names throughout the country,[29] and medial replacement of /ð/ with /d/ is found throughout fifteenth-century England.[30] If this is a dialect feature, it is not easily localizable; the interchange of d and þ is

Romance: The Limits of Editing, the Limits of Criticism", in *Medieval Literature: Texts and Interpretation*, ed. Tim William Machan (Binghamton, New York, 1991), 91-104.

[25] *Lybeaus Desconus*, 37.

[26] *Sir Launfal*, 9.

[27] *Handbook of Middle English Grammar: Phonology*, trans. and rev. Eugene Crook. (The Hague, 1974), 186-187.

[28] A. J. Bliss, "The Spelling of *Sir Launfal*", *Anglia* 75 (1957), 275-289.

[29] The shift of initial /θ/ to /ð/ and the shift of /ð/ in any position to /d/ are common in place names (e.g., Thorndenn > Dornden and Scathebury > Scadbury).

[30] Samuels, "Kent and the Low Countries", 11.

found in a fragment of the *The Wars of Alexander* from the north as well as in the more southern Caligula.[31] I would like to make the tentative suggestion that this is a type of orthographic shorthand, although more study of the phenomenon is necessary.

Some texts display an East Midlands pedigree.[32] For example, *Emaré* is written in a northeast Midlands dialect with significant northern forms and northern vocabulary that lead Rickert to place the text between the Humber and Karesborough in mid-Yorkshire.[33] Frances McSparran observes a SE Midland dialect,[34] and Ralph Hanna and David Lawton are more specific and argue that at least the *Siege of Jerusalem* demonstrates an East Anglian provenance, although they disagree with McSparran's SE Midland attribution of *Octavian Imperator*. Their conclusions are worth quoting in full:

> However, our experience with the forms of *The Siege* would suggest that Samuels[35] has in fact identified a substrate peculiar to the transmission of *Octavian*, perhaps a London-Essex text: within *The Siege* scarcely any form occurs that one would associate with Kent. Here rather, the scribe seems to have assimilated, probably from relicts in his archetype, a melange of various East Anglian forms, focussed in no single area. Various aspects of the scribe's language show passing affinities with forms peculiar to (1) the western juncture of Norfolk and Suffolk with Ely and Cambs., (2) to King's Lynn and immediately adjacent Lincs. (congruent with Tattershall provenance); (3) to East Suffolk; and (4) to south central Norfolk. These may provide nothing more than evidence of protracted copying history in this area.[36]

My own observations of the manuscript's dialect show even further variation. To give one example, the ending -us in unstressed final syllables common in the West Midlands[37] appears in, for instance, the *Charter of*

---

[31] The fragment was transcribed by Neil Ker in 1941 and is printed in Thorlac Turville-Petre, "A Lost Alliterative Alexander Romance", *Review of English Studies* 30 (1979), 306-307.

[32] See Henry H. Gibbs, ed., *Cheualere Assigne* (London, 1868), xv-xviii.

[33] Rickert, *Emaré*, xviii.

[34] McSparran, *Octovoian Imperator*, xxviii.

[35] Samuels was aware of this: "Apart from the feature under discussion and a few others, the forms are preponderantly those of the copyist's originals and not his own. It would require exhaustive study of the whole MS to decide whether there are sufficient traits common to the whole of it for the latest stratum of its language to be localized", "Kent and the Low Countries", 19n.73.

[36] *The Siege of Jerusalem*, xxvi.

[37] See *LALME* dot maps 642 and 958.

Christ ("wordus", l.4) and *Quindecim Signa* ("fowlus and fyshchus", l.5). These forms are used inconsistently and infrequently; much more often, we find -es and -s in plural (occasionally, we find -n, -is, and -ys). As a result, there are markers in this manuscript that point to Yorkshire, the West Midlands, the East Midlands, East Anglia, and Kent, and it would only be possible to distinguish these strata and examine their relationships if Caligula's texts were compared to the manuscript's exemplars; however, no stemma has yet uncovered a surviving text or manuscript that may be a direct exemplar of Caligula.

For now, we can argue that the manuscript's dialectical variation uncovers the scribe's textual conservatism; he was more of a transcriber than an editor (with the important exception of *The Siege of Jerusalem*, discussed by Hanna and Lawton)[38] and the resulting dialectical strata of his texts suggest that his exemplars were themselves late copies of older texts that were rewritten and re-transcribed over generations in many different parts of the country. Many texts in Caligula have modernised grammatical forms; the use of "me seyn" in *Eustache* (l.130) is replaced with "mene seyn", thus transforming the verb to active voice and introducing an agent that provides the text with a certain level of proverbial authority. This is what one would expect from a mid-fifteenth century manuscript preserving old texts such as *Susan*, which was probably written over a century before Caligula was compiled. Thus while the scribe's dialect is indeterminate, we can say that he copied his texts carefully. Any scribal emendation, therefore, is significant.

We also find *Tundale* in National Library Scotland Advocates 19.3.1 (ff.98-157v), otherwise known as the Heege manuscript.[39] This book can be connected to the Sherbrooks of Nottinghamshire, whose absence from local records before the sixteenth century suggests that they were not of particular importance and possibly not landowners,[40] although by the mid-fifteenth century at least one member of the family, Cuthbert Sherbrook, had amassed a large number of books.[41] By 1551, Thomas Sherbrook was

---

[38] *The Siege of Jerusalem*, .lx-lxviii.

[39] The manuscript has been reproduced and introduced in Phillipa Hardman, *The Heege Manuscript: A Facsimile of National Library of Scotland MS Advocates 19.3.1* (Leeds, 2000).

[40] See Thorlac Turville-Petre, "Some Medieval Manuscripts in the Northeast Midlands", *Manuscripts and Readers in Fifteenth-Century England*, ed. Derek Pearsall (Cambridge, 1983), 125-141, at 139-40.

[41] At the beginning of the twentieth century a cache of "between forty and fifty volumes" was found in Oxton hall; the "bulk" of these books contained Cuthbert's signature—see ibid., 136.

important enough that his move to Oxton from Tibshelf was recorded, and by 1600 the family was wealthy enough to begin buying land in the area.[42]

Although apparently not an important family in the fifteenth century, the Sherbrooks were presumably well-off enough to afford the costs involved in producing this book, including hiring and paying the book's main scribe, Richard Heege, who was not a member of the Sherbrook family, and whose family name identifies him with an area not far from the Sherbrooks' home—the village of Heage, in eastern Derbyshire, near the Nottinghamshire border.[43] Like Caligula, the Heege manuscript is made entirely of paper and is of a similar size (it is about 210mm x 145mm; Caligula is about 210mm x 140mm). It was possibly produced a generation or two after Caligula.[44] The manuscript's collation[45] suggests that it comprises a series of individual booklets, which Phillipa Hardman argues are constructed upon separate but interconnected themes.[46] The manuscript preserves three texts often identified as romances: *Sir Gowther*, *Sir Amadace*, and *Sir Isumbras (*ff.48-56v), which is also preserved in Caligula (ff.130-134). The Heege manuscript and Caligula also share *Trentale Sancti Gregorii*, "Urbanitatis", and *Tundale*.

Heege's copy of *Tundale* is particularly interesting because the scribe (or his exemplar) has added a skeptical couplet to the end of the text:

> Be it trwe, or be it fals,
> Hyt is as the coopy was. (2382-2383)

Here the scribe seems to doubt the truth-value of the text while asserting the accuracy of his job of copying the work. We cannot conclude from this that the scribe and his readers took it as a work of sheer fancy to be read for entertainment alone, since it is preceded by a form of indulgence and followed by William Lychefelde's *Complaint of God*, a popular Middle

---

[42] Phillipa Hardman, *The Heege Manuscript*, 1.

[43] The bulk of the Heege manuscript was written in a single secretary hand with Anglicana features. Three secondary scribes collaborated with Heege: John Hawghton (who signs his name on f.95v and f.96) and two who have not been identified, while other hands contributed items throughout the manuscript and wrote marginalia on several folios. See Hardman, *The Heege Manuscript*, 6. Hardman calls the hands of all the scribes "typical late-fifteenth-century mixed cursive hands" (ibid., 6).

[44] Hardman suggests that the manuscript was produced before 1478 on the basis of one of its watermarks (Briquet 22—see ibid., 1).

[45] The manuscript is made up of a total of thirteen booklets of varying length—see ibid., 4-5.

[46] Ibid.,14-39.

English poem in 8-line stanzas that underscores the inevitable sinfulness of humanity and the need for contrition. As a companion piece, *Tundale's* horrific depictions of torturous punishments for the wicked is appropriate if it is read as a stern warning. Perhaps here Heege embraces the moral sentence of this work while maintaining a healthy skepticism about its literal veracity.

Tundale also survives on ff.150-184v of BL Royal 17.B.XLIII, another manuscript known for its romance content (in this case *Sir Gowther* on ff.116-131v), although the majority of the codex is devoted to *Sir Mandeville's Travels* in a booklet currently at the beginning of the book. The book ends with a Middle English carol produced in an early sixteenth-century hand that was probably considered an appropriate addition to visions of the afterlife. A rubricated date of 1451 on f.184 suggests that the manuscript was probably completed in this year, making it the oldest extant manuscript of the text. The late date of this manuscript is probably an accident of history, since the language of *Tundale* in all of its manuscript copies seems much older.

While M. C. Seymour has suggested that the book as it stands was originally four separate manuscripts,[47] I agree with Rodney Mearns that similarities and symmetrical patterns in the book suggest that the booklets containing *Tundale* and *Mandeville* were probably produced together,[48] and I would also suggest that *Gowther*, although written in a significantly more decorative hand, was also produced in tandem and was part of the scribe's larger plan for the book. The closest thematic and scribal link provided by the manuscript is *St. Patrick's Purgatory*, here in prose on ff.133-148v. Preceding this work is an illustration of a "nimbed bishop in the act of benediction, surrounded by fiends and souls in torment";[49] an equally appropriate illustration, seemingly of Tundale's guardian angel, is identified by Mearns as a work of the same artist; this follows the prose *Purgatory* and precedes *Tundale* on f.149v.[50] The presence of these illustrations on the versos of otherwise blank inserts between booklets, the iconographic and stylistic similarities, and the relationship of the illustrations to the texts provide ample evidence to suggest that at least *Purgatory* and *Tundale* were produced alongside one another, probably for the same or similar audience.

---

[47] M. C. Seymour, "The English Manuscripts of *Mandeville's Travels*", *Edinburgh Bibliographical Society* 4.5 (1966), 185-186.

[48] *The Vision of Tundale* 54-56.

[49] G. F. Warner and J. P. Gilson, *British Museum Catalogue of Western Manuscripts in the Old Royal and King's Collections* (London, 1921), II.233-234.

[50] *The Vision of Tundale*, 54-55.

Contemporaneous to Caligula and Royal is Tokyo, Takamiya 32, *olim* Boies Penrose Library MS 6, where an imperfect copy of *Tundale* is to be found on ff.166v-175v immediately after *Partenope of Blois* (ff.164-165v), another Middle English romance focusing on the adventures of a central knight-protagonist.[51] This book was probably produced in the 1450s somewhere in the East Midlands, and is unique in preserving Gower's *Confessio Amantis* alongside part of Chaucer's *Canterbury Tales*. *Tundale* is preserved near the end of the manuscript (on ff.166v-175) and is followed by *The Gast of Gy*, which ends imperfectly in this manuscript (only 37 lines survive on f.175v).

The scribe's attitude to the text is made enigmatic by its vague title, probably unique to the scribe: "her be gynniht a good storie of Tundale" (f.166v). We are left to imagine whether it is good because it is moral, entertaining, true, miraculous, or all of the above. Spaces for three-line capitals remain, which neatly divide major movements in the narrative from Tundale's journey through Hell to his eventual contemplation of Lucifer's punishment. As Mearns notes, the scribe's "preoccupation is with the structure of the work",[52] and his editorial alterations to the text—mostly additions—seem to serve to enliven the story.

The final manuscript preserving this text is even more fragmentary than the Royal manuscript and is also the most enigmatic: Bodleian Ashmole 1491, which preserves a fragment of Tundale on six scorched leaves bound at the back of the book alongside previously unrelated materials. Since the text is so fragmentary, no comprehensive analysis of its presumed provenance, scribal hand, or even date has been made, although I agree with Rodney Mearns that it is a fifteenth-century production. The vellum leaves preserve the text in double columns of 35 lines and appear to have been produced with a careful visual presentation

---

[51] The manuscript has been analyzed by Helen Cooper, "Textual Variation and the Alliterative Tradition: Canterbury Tales I.2602–2619, the D Group and Takamiya MS 32", in Takami Matsuda, Richard Linenthal, and John Scahill, eds., *The Medieval Book and a Modern Collector: Essays in Honour of Toshiyuki Takamiya* (Cambridge, 2004) 71–80 and by A. S. G. Edwards in "Gower in the Delamere Chaucer Manuscript", in ibid. 81–86. However, the fullest study of the manuscript and its copy of *Tundale* can be found in Mearns, *The Vision of Tundale*, 49-50. See also Gisela Guddat-Figge, *Catalogue of Manuscripts Containing Middle English Romances* (Munich, 1976), 112-113, who bases her analysis mostly on the Manly-Rickert edition of the *Canterbury Tales*—see J. M. Manly and E. Rickert, *The Text of the Canterbury Tales*, 8 vols. (Chicago, 1940), I.108-116.

[52] *The Vision of Tundale*, 50.

in mind, which have been discussed at length by Mearns.[53] Most important for our present concern is his observation that most textual emendation in this book seems to have been made out of "a primary concern with the visual impact of the MS at the expense, to some extent, of the subject-matter".[54] This sacrifice of textual content for visual presentation is not unusual in the fifteenth century. Similarly common for the era is the fact that the text seems to have been catered to a reading audience specifically in a codex that is a sophisticated and self-conscious work of art produced according to a meticulous plan. Further, the fact that the fragment also preserves portions of a spiritual text in French confirms that this audience was erudite as well as devout.

What is immediately striking about these manuscripts is that, with the exception of the fragmentary Ashmole, they all preserve Middle English romances. The combination of orthodox religious material with romances is often ignored by scholars, yet many fifteenth-century romance manuscripts cater to religious and romantic impulses within its readership without the threat of heterodoxy or theological subversion. It is rarely noticed that Lollard materials and Middle English romances do not survive alongside one another in medieval manuscripts, although both corpuses are substantial and account for a large portion of vernacular scribal energy in the fourteenth and fifteenth centuries. Furthermore, *Tundale* shares with the Middle English romances not only spiritual conservativism, but also many stylistic and thematic traits, such as an interest in the supernatural or out-of-the-ordinary, a narrative focus on a singular person who experiences *aventures* towards a final happy resolution, and a vivid description of supernatural images and imaginative wonders.

Also immediately apparent is the time-gap between the Latin *Visio Tnugdali* and manuscript copies of its Middle English translation. While it is impossible to date the authorship of the Middle English text, it seems likely a product of the fourteenth century, leaving us to wonder why a gap of over a century separates the Latin text with its identifiable English readership. Perhaps this is an accident of history resulting from the increased number of manuscripts in the fifteenth-century and bad luck destroying earlier copies; perhaps it demonstrates a long delay between the transmission of miraculous rumors in Cistercian circles and a vernacular audience. In any case, the broad historical range demonstrates the durability of these stories as texts to capture the imagination of readers, even if their approach to the story differed. To give a modern analogue,

[53] Ibid., 50-52.
[54] Ibid., 51.

Jane Austen's novels have yet to survive the test of time that *Tundale* did by the time that these manuscript copies were made.

The inevitability of God's reckoning so central to *Tundale* is, perhaps unsurprisingly, a feature of many religious texts that survive alongside it in the Middle English tradition. Thus, the text is wholly orthodox in theme even if it deviates from orthodox doctrine in describing the horrors of Hell. Scribes such as Heege may have been aware of its thematic value even if they were skeptical of its veracity, and it is thus on a thematic and ideological level that they and their readers responded to the text.

It is tempting to consider *Tundale* a work of popular spirituality, yet the presence of Ashmole should remind us that medievalists, particularly codicologists, constantly work with fragmentary data, and we can never know what has been lost. The fragmentary survival of Ashmole may be as much an accident of history as is the survival of four much better copies of the text in a seemingly more popularized context; quite possibly, the fifteenth-century saw more renditions of *Tundale* into English for erudite audiences that have been lost over time, leading us with the distorted view of the text as a popular one.

However, the fact of the four less sophisticated versions of the text remains, and we are left with the impression that, regardless of its other appealing qualities, *Tundale* contributed to the textual expression of popular religious sentiment in Middle English throughout the fifteenth century. The fact that this popularizing and secularizing impetus is imposed on a twelfth century Cistercian text has often been acknowledged but never fully analyzed. In what follows I would like to read the Caligula compiler's editing as an attempt to guide his readership towards a "proper" theological understanding of the text's contents and implications. His revisions to the text, therefore, are what I would like to call forms of "theological rhetoric".

I have chosen Caligula because it gives the fullest version of the text, yet its scribe has altered it enough to suggest that he was concerned with the ethical, spiritual, and ideological implications of the text and with his intended readership's presumed lack of theological and epistemological grounding to understand the text appropriately—i.e., within the limits of accepted Christian orthodoxy.

Rhetorical Variants in the Caligula Text

The most immediately obvious scribal contribution of significance to Caligula's copy of *Tundale* is his decision to place it immediately after *Owayne Miles* (on ff.91v-95). This is an obvious companion piece to

Tundale as an Irish narration of a man who descends into the netherworld, and it inspires an interest in how one's life determines one's afterlife with a detailed description of Purgatory and Hell. The decision of Caligula's scribe to copy these two texts beside one another seems largely motivated by his interest to inspire concern for the afterlife amongst his audience. This theme is repeated throughout the manuscript; the religious lyrics "Make Amendes" and the prose form of confession following it (f.69 and ff.69v-70), the poetic hymn "Amen for Charite" immediately preceding *Owayne Miles* (f.91), and the lamentation "Myn Owene Woo" (f.108-108v) approach the theme of penance from different perspectives with the same obvious ethical conclusion: repent lest ye suffer. The form of suffering implied by these texts is made explicit in the two Irish visionary texts.

Besides minor scribal variants that I shall not consider here, there are a few variations in the Caligula text that appear to be moments of rhetorical emendation. The lengthiest of these is a lacuna at ll.1305-1372. In these lines an angel describes to Tundale how God punishes Lucifer and the state of his pit, where he suffers for all eternity in complete darkness. Its emphasis is both on the horror of Lucifer's punishment and the superlative quality of the image itself ("Soo ugly was þat lothely syght...So orryble he was ferede", 1317; 1319). It is an informative and detailed passage, giving Lucifer's shape and size in numbers of cubits, and describing the texture of his body and nails.

At first glance, it is unclear why the Caligula scribe has not copied this section. The omission serves no clear structural or literary purpose; although it is unnecessary, the same could be said for much of the rest of the poem's highly detailed descriptions of sinners' punishments. On the surface, this passage is quite similar to the rest of the poem, and in many ways it could be seen as its crown jewel—the description of the greatest punishment for the greatest sinner of all, Satan himself.

I believe the centrality and intensity of this passage is ironically why the scribe decided to remove it; elsewhere we read about *human* punishment that easily translates to the reader, but the devil's pains serve as a poor example for what happens to human beings, and brings up uncomfortable questions about the relationship between angels and humans. Also, and more disastrously, describing the devil's suffering could inspire sympathy for him (such as Blake's famous response to Milton), and contradict the fear of him and of Hell that the text attempts to inspire in its readership. Without it, the scribe avoids such questions and returns the focus to God's punishment on humans, which is the central theme of this and other texts in Caligula.

Caligula's copy of *Tundale* similarly avoids all eschatological considerations and questions of the relationship between the metaphysical and physical worlds by sidestepping Satan's punishment and devoting more attention to the fates of human sinners in the afterlife. The very physical descriptions of souls in *Tundale* and *Owayne Miles* confuses a distinction between body and soul, perhaps because the authors of these works are working within an Aristotelean notion of the soul as an inseparable part of the body or as the cause of physical sensations that was later developed yet unresolved by Thomas Aquinas (*Summa Theologiae* pt.1.75, art.1-3), and was further refined by Réne Descartes into the body/soul dualism doctrine that has been questioned by modern evolutionary psychology.[55] The body/soul problem is thoroughly ignored in *Owayne Miles*, which makes no distinction whatsoever between the physical body and the metaphysical soul and suggests that the afterlife is a real physical place that one can walk into. *Tundale's* more didactic and less philosophical focus is also compatible with the various penitential texts in Caligula discussed above, which never dwell on eschatology or complex theological questions. The omission of this passage, then, is borne out of a scribal urge to focus on the fate of the human soul and avoid distractions, such as the nature of angels or the relationship between the human body, the soul, and angels.

Rodney Mearns has hinted at the scribe's didactic concerns by noting that "the dual considerations of visual presentation and suitability to a particular readership were influencing his [i.e., Caligula-scribe's] principles of selection and rejection",[56] although it is unlikely that the scribe chose to remove the passage on Lucifer to protect the squeamish, since other gruesome passages are left unedited. Other omissions were probably not made for this reason, either, such as the loss of two lines after a scene in which souls were burned on a lake of ice and then bitten by "edderes & oþer vermyne", which would then trace through the sinners'

> hede, fote, bakke & syde,
> Thorow legges & armes gan þey glyde;
> [Thorghw wombe and brest they cam crepande,
> And that made hem well sore gretande.] (941-944)

Possibly, Caligula's scribe felt that this final couplet was expendable because it added little to the text's meaning; it seems unlikely that this was

[55] For an accessible study of this issue, see Steven Pinker, *The Blank Slate: The Modern Denial of Human Nature* (New York, 2002).

[56] *Vision of Tundale*, 47.

too gruesome for the book's readers, since the visual details before it are left intact.

Another small omission of gruesome detail is found after a lengthy description of how smiths in Hell punish souls (and Tundale, at this point, whom they have taken) with the tools of their trade. A couplet in all other manuscript copies reads:

> Hem þouȝt þey were not smethede inowe,
> But throwe a fire efte they hem drowe; (1111-1112)

Perhaps the scribe felt that they had received punishment enough and worried that readers would begin to sympathise with the damned, so he left these lines out. Also, its omission does little to the general impetus of the text, and allows the story to flow more quickly.

Another more compelling redundancy is removed at lines 1675 and 1678, after a lengthy description of the splendid hall where Tundale sees Conor O'Brien ("Cantaber") and Donough McCarthy ("Donatus") in bliss. Tundale is shocked at the sight, and mentions that these two Irish kings were cruel leaders: "In hem was lytull mercy sene" (1610). The angel then explains that the kings repented before death, thus reinforcing the general theme of the poem and the purpose of Tundale's descent. These kings become hyperbolic examples for Tundale to learn the benefits of contrition (and, in turn, they are examples for *Tundale's* readership), so their open-aired hall is beautiful and made of riches, including the carbuncle roof, a floor of precious stones, and walls of gold: "Hyt semed a hows of grette coste" (1656).

Because of the importance of these kings as examples of the power of repentance, it is unsurprising that Caligula's scribe has not truncated this passage. Yet we find an omission of two lines after this scene, when the attention turns to King Cormac MacCarthy, (Cormake), who was an important Irish king and benefactor to the Irish Benedictine order and the foundation at Regensburg. This context was probably unknown to Caligula's scribe and its audience, and, even if they did know about an Irish king who had been dead for three centuries and who had ties to a distant Bavarian monastery, it is doubtful that these historical facts would have been as interesting or as important to them as the theological implications of penance and the bliss of Heaven. Thus it is easy to understand why the scribe shortened his text at this moment by removing two lines:

Tundale stode ner & toke good kepe	Tundale stode ner & toke good kepe
And byhelde longe þat grete worshype	And byhelde longe þat grete worshype
That men þer to Cormake dyde	That men þer to Cormake dyde
That somtyme her was [hys lord] kydde.	That somtyme her was kydde.
[For he was symtyme wyt hym of meyne]	
Ther ferly of þat syȝte hadde he.	Ther ferly of þat syȝte hadde he.
Prestes & dekones þer wer manye.	Prestes & dekones þer wer manye,
[Before hym a fayre company.]	
(Mearns ed., 1671-1678)	(Caligula, 1671-1678)

Here the omission of lines 1675 and 1678 allows 1676 and 1677 to become a couplet without losing the text's sense or rhetorical force. The scribe is able to maintain the text's relevance to its new audience while de-emphasizing part of its history less immediately important or relevant to their own life experiences and concerns.

While Mearns suggests that the scribe omits passages that castigate "those who are in positions of authority for being preoccupied with power",[57] I would suggest instead that there is no consistent purpose behind Caligula's various omissions, and each needs to be examined on its own as the scribe's effort to adapt his text to a new audience and to make it appeal to their presumed interests and theological knowledge. The efforts of this scribe to shift the audience's focus to different points of the text are reminiscent of the medieval tradition of amplifying and abbreviating sources to fit new rhetorical circumstances. The scribe's efforts in producing his text seem to have expanded beyond reproduction and into the realm of rhetorical emendation.

Conclusion

Contrary to the popular perception of medieval dogma as a fixed cultural mandate from a central authority, the form of popular Christian faith that *Tundale* communicates in its various manuscript versions is a constantly mutable and diverse set of ideas within the structure of orthodox belief. While it was, by the fifteenth century, neither subversive nor new, scribes still re-organized it according to their concern for how it may lead readers astray. This type of scribal editing, which I have here

[57] Ibid., 47.

called "theological rhetoric", is a widespread phenomenon in medieval textual culture, although it has gone ignored as editors focus more on presumably subconscious emendations from eyeskip, dialectical translation, and the alteration of rare words for more common ones.

The textual diversity of *Tundale* suggests that scribal emendation is a more complex and self-conscious activity of adjusting texts to new audiences, whether for visual layout as in Ashmole, or as an expression of moral truth (but perhaps also a factual untruth) in Heege, or to steer readers towards a proper theological understanding of the text, as I have argued for Caligula. Such textual alterations demand closer attention because they are part of the social history of how texts' ideas spread and were manipulated by scribes asserting their own morality. It is not news to learn that a complex network of editors, publishers, and booksellers stands between authors and their intended audiences today, and it should be no surprise that, in the middle ages, an added layer of scribes would contribute to the theological and intellectual discourse of their day with a silent, unexpected, and possibly unwanted filter.

Bibliography

St. Augustine. *Tractates on the Gospel of John 55-111*, ed. John W. Rettig (New York, 1994).

Barolini, Teodolinda. *Dante and the Origins of Italian Literary Culture* (New York, 2006).

Bliss, A. J., ed. *Sir Launfal* (London, 1960).

Bliss, A. J. "The Spelling of *Sir Launfal*", *Anglia* 75 (1957), 275-289..

Boswell, C. S. *An Irish Precursor of Dante: A Study of the Vision of Heaven and Hell Ascribed to the Eighth Century Irish Saint Adamnán* (London, 1908).

Brown, Carleton, ed. *Religious Lyrics of the Fifteenth Century* (Oxford, 1939).

Cooper, Helen. "Textual Variation and the Alliterative Tradition: Canterbury Tales I.2602–2619, the D Group and Takamiya MS 32", in *The Medieval Book and a Modern Collector: Essays in Honour of Toshiyuki Takamiya*, Takami Matsuda, Richard Linenthal, and John Scahill, eds., (Cambridge, 2004), 71-80.

D'Ancona, Alessandro., *I precursori di Dante* (Florence, 1874).

Easting, Robert. "The Date and Dedication of the Tractatus de Purgatorio Sancti Patricii", *Speculum* 53 (1978), 778-83.

—. ed. *St. Patrick's Purgatory* (Oxford, 1991).

Edwards, A. S. G., "Gower in the Delamere Chaucer Manuscript", in *The Medieval Book and a Modern Collector: Essays in Honour of Toshiyuki Takamiya*, Takami Matsuda, Richard Linenthal, and John Scahill, eds., (Cambridge, 2004), 81–86.

—. "Middle English Romance: The Limits of Editing, the Limits of Criticism", in *Medieval Literature: Texts and Interpretation*, ed. Tim William Machan (Binghamton, New York, 1991), 91-104.

Emerson, Jan. *Imagining Heaven in the Middle Ages: A Book of Essays* (London, 2000).

Gardner, Eileen. *Visions of Heaven and Hell Before Dante* (New York, 2008).

Gibbs, Henry H., ed. *Cheualere Assigne* (London, 1868).

Glauning, O., ed. *Lydgate's Minor Poems* (Oxford, 1900).

Guddat-Figge, Gisela. *Catalogue of Manuscripts Containing Middle English Romances* (Munich, 1976).

Hanna, Ralph and David Lawton, eds. *The Siege of Jerusalem* (Oxford, 2003).

Harbison, Peter. "Twelfth and Thirteenth Century Irish Stonemasons in Regensburg (Bavaria) and the End of the 'School of the West' in Connacht", *Studies: An Irish Quarterly Review* 64 (1975), 333-346.

Hardman, Phillipa. *The Heege Manuscript: A Facsimile of National Library of Scotland MS Advocates 19.3.1* (Leeds, 2000).

Harvey, P. D. A. *Mappa Mundi: The Hereford World Map* (London, 1996).

Jordan, Richard. *Handbook of Middle English Grammar: Phonology*, trans. and rev. Eugene Crook. (The Hague, 1974).

Kane, George and E. T. Donaldson, eds. *Piers Plowman, The B Version*, 2nd ed. (London, 1988), 128-220.

Le Goff, Jacques. *The Birth of Purgatory* trans. Arnold Goldhammer (Chicago, 1986).

MacCracken, H. N., ed. *The Minor Poems of John Lydgate* vol. 1 (Oxford, 1911).

Manly, J. M. and E. Rickert. *The Text of the Canterbury Tales*, 8 vols. (Chicago, 1940).

Matsuda, Takami. "*Sir Gawain and the Green Knight* and St. Patrick's Purgatory", *English Studies* 88 (2007), 497-505.

McSparran, Frances, ed. *Octovian Imperator* (Heidelberg, 1979).

Mearns, Rodney, ed. *The Vision of Tundale Edited from BL MS Cotton Caligula A.II*, (Heidelberg, 1985).

Mills, M. "The Composition and Style of the "Southern" *Octavian*, *Sir Launfal*, and *Libeaus Desconus*', *Medium Aevum* 31 (1962), 88-109.

—. *Lybeaus Desconus* (London. 1969).
Morgan, Alison. *Dante and the Medieval Other World* (Cambridge, 1990).
Rickert, Edith. *Emaré* (London, 1908).
Samuels, M. L. "Kent and the Low Countries: Some Linguistic Evidence", in *Edinburgh Studies in English and Scots*, ed. A. J. Aitken, Angus McIntosh and Hermann Palsson (London, 1971), 3-26.
Seymour, M. C. "The English Manuscripts of *Mandeville's Travels*", *Edinburgh Bibliographical Society* 4.5 (1966), 185-186.
Turville-Petre, Thorlac. "A Lost Alliterative Alexander Romance", *Review of English Studies* 30 (1979), 306-307.
—. "The Persecution of Elizabeth Swillington by Ralph, Lord Cromwell", *Nottingham Medieval Studies* 42 (1998) 174-87.
—. "Some Medieval Manuscripts in the Northeast Midlands", *Manuscripts and Readers in Fifteenth-Century England*, ed. Derek Pearsall (Cambridge, 1983), 125-141.
Warner, G. F. and J. P. Gilson, *British Museum Catalogue of Western Manuscripts in the Old Royal and King's Collections*, 4 vols. (London, 1921).
Woolf, Rosemary. *The English Religious Lyric in the Middle Ages* (Oxford, 1968).
Zaleski, Carol. *Otherworld Journeys: Accounts of Near-Death Experience in Medieval and Modern Times* (Oxford, 1988).

CHAPTER EIGHT

PLAY TIME: PICTURING SEASONAL GAMES IN THE SIXTEENTH CENTURY

AMY ORROCK

A close bond existed between time and daily life in medieval Europe. At the heart of this were the changing seasons, which dictated the constant round of labours necessary for survival—from harrowing the soil and sowing seeds in spring to harvesting crops and slaughtering animals in autumn. These daily rituals are brilliantly captured in the pages of illuminated manuscripts, with the most detailed depictions found in the calendars of Books of Hours and Breviaries produced in France and the Netherlands in the fifteenth and sixteenth centuries (fig. 1). In addition to charting the everyday tasks and duties of the peasants, many illuminated manuscripts also offer us glimpses of how leisure time was spent. The images of seasonal games, religious festivals and carnival revelries found in the manuscript calendars all suggest that during this period time could be measured through seasonal play as well as seasonal labour. A number of Netherlandish manuscripts even include separate borders in which a "game of the month" cycle can be identified, complementing the established cycle of seasonal labours. Depictions of a wide variety of popular seasonal pastimes can also be found in the panel paintings of Pieter Bruegel the Elder (c.1525-1569). Bruegel's scenes may be on a far grander scale than the manuscript pages, but their playful iconography is remarkably similar, with small figures skating and sledding in winter or busily engaged in archery and bowling tournaments in summer. Bruegel scholars have drawn attention to these similarities and manuscript calendars have proved particularly useful in decoding the iconography of Bruegel's great series of seasonal panels known as the *Months*. Yet in Bruegel's most sophisticated representation of games, *Children's Games* (Vienna, Kunsthistoriches Museum), the temporal aspect of game playing found in illuminated manuscripts is entirely absent. Beginning with the development of game-calendars in French manuscripts of the fifteenth century, and then moving on to consider the notion of the "game of the month" in relation to Bruegel's oeuvre, this essay seeks to explore the links between games and time in the later medieval period by unraveling some of the more complex ideas contained within these playful images.

Calendar Games in French Manuscripts

The tradition of using leisure activities in addition to labours to represent the changing seasons first emerged in fifteenth-century France, with the influential *Très riches heures* of Jean, duc de Berry (Chantilly, Musée Condé MS. 65). As the opening section of a Book of Hours, the calendar had long been an important and favoured site for decoration. Functioning as the introduction to a manuscript, the calendar set the tone and needed to entice the reader to continue. The calendar was also the place where the sacred was married with the secular. Appropriate forms of decoration could thus include everything from images of saints and Biblical events to zodiacal signs and seasonal tasks. The sumptuous calendar of the *Très riches heures* offered more than just a particularly lavish version of this iconography. In a new development, the calendar personalised the traditional occupations of the month by picturing them from the perspective of the duke, on his land and often within sight of his properties. The monthly scenes offer an inventory of the duke's material wealth: a splendid banqueting hall is the backdrop for the January miniature; in the miniature for July reaping and shearing takes place before the Château du Clain in Poitiers, built by the Duke toward the end of the fourteenth century; in August the Château d'Etampes appears in the background.[1] In addition to detailing the duke's considerable wealth and holdings, the *Très riches heures* calendar miniatures also inventory the duke's privileged lifestyle by relegating the traditional peasant labours to the background and foregrounding aristocratic leisure activities instead. In

[1] For reproductions of the calendar pages, see Raymond Cazelles and Johannes Rathofer, *Illuminations of Heaven and Earth: The Glories of the Très Riches Heures du Duc de Berry* (New York, 1988). The Limbourg brothers did not complete the manuscript and several later hands have been identified in the calendar miniatures, including that of Jean Colombe and a so-called "intermediate painter". Debates about the precise dating of each of the calendar miniatures affect our understanding of details in the iconography, such as the identities of some of the noblemen and castles pictured.

the January miniature the duke himself is pictured seated at a banqueting table, surrounded by his servants and pets, in May noblemen and ladies ride in the woods, and in August the aristocratic leisure activity of hawking takes place in the foreground while in the background peasants cut corn and swim in the river. The distinction made in the *Très riches heures* between aristocratic leisure and peasant labour served ultimately to underline the status of the duke, and the success of the strategy of picturing such "divergent milieus" was confirmed by the fact that a number of scenes of aristocratic leisure introduced in the manuscript were to become standard fare in later calendar iconography.[2] The miniatures of the *Très riches heures* influenced a generation of Netherlandish artists, most notably the important group of illuminators behind the famous *Grimani Breviary* (Venice, Biblioteca Nazionale Marciana) who may have come across the French manuscript at the court of Margaret of Austria in Mechelen.[3]

The landed classes were not the only people to be depicted enjoying seasonal leisure pursuits in manuscript calendars. Numerous fifteenth- and sixteenth-century manuscripts attest to a growing interest in picturing the games and pastimes of the middle and lower classes. One such example is the calendar of a manuscript known as the *Hours of Adélaïde de Savoie* (Chantilly, Musée Condé MS Lat.1362). With each month represented over two pages, this calendar includes illustrations of many typical seasonal labours—we find peasants harvesting hay and corn in June and July, preparing wine barrels in August, picking fruit in October and killing a pig in November. But interspersed with these traditional agricultural tasks are a number of more pleasurable pastimes. In February, Valentine's Day lovers play a game of *petit papiers*, a courting game in which men and women choose slips of paper; in May, figures are pictured dancing; in June they play *jeu de paume*, an early form of tennis; in July a water joust is staged and in December the whole town participates in a snowball fight. While we do not know the identity of the original owner of the manuscript (it takes its name from its seventeenth-century owner, Adélaïde, Duchess

---

[2] For discussions on the social implications of the *Très riches heures* iconography, see Erwin Panofsky, *Early Netherlandish Painting: Its Origins and Character,* 2 vols. (Cambridge, Mass., 1964), 66 and J. J. G. Alexander, "*Labeur* and *Paresse*: Ideological Representations of Medieval Peasant Labour", *Art Bulletin* 72 (1990), 443-452.

[3] Thomas Kren and Scott McKendrick, *Illuminating the Renaissance: The Triumph of Flemish Manuscript Painting in Europe* (Los Angeles, 2003), no. 126.

of Burgundy), it has been proposed that these urban, upper-middle class games might reflect the interests and milieu of the book's patron.[4]

Calendars of games appear in two other French books; a printed Book of Hours now in the Pierpont Morgan Library (New York, PML 572) and a little-known hand illustrated Book of Hours in the Bodleian Library (Oxford, Douce MS 135).[5] The close relationship between these two books has never before been recognised, but when their calendar pages are placed side-by-side the similarities are clear (fig. 2 and fig. 3).[6] Both feature highly decorative twelve-page calendars, which are distinguished by large lower borders containing images of games and pastimes—the example pictured presents a version of a popular guessing game known as *frappé main*, "hot hand" or "hot cockles".[7] Produced in Paris by Philippe Pigouchet for Simon Vostre, the printed *Horae* is dated 1497, and thus probably predates the sixteenth-century Douce manuscript.[8] The designer of the metal cuts for the Parisian *Horae* has been identified as the Master of Anne de Bretagne, a well-known and versatile artist who had a shop that designed images for books, drew cartoons for stained glass and tapestries, and produced murals and manuscripts.[9] Although named after a

---

[4] Janine Bouissounouse, *Jeux et travaux d'apres un livre d'heures du XV siecle* (1925; 2nd edn., Geneva, 1977).

[5] In their entry on Douce 135, Pächt and Alexander note the resemblance between it and a manuscript now in Paris (BNF, MS Lat. 9446) but the resemblance between Douce 135 and PML 572 has not yet been noted. See Otto Pächt and Jonathan J. G. Alexander, *Illuminated Manuscripts in the Bodleian Library Oxford* (Oxford, 1966), no. 851.

[6] For an image of the September and October pages of PML 572, see Roger Wieck, *Painted Prayers: The Book of Hours in Medieval and Renaissance Art* (New York, 1997), 33. For reproductions of some of the calendar pages from Douce 135 see Bridget Henisch, *The Medieval Calendar Year* (Pennsylvania, 1999), colour plate 7-14 and figs. 4-9, 5-8, and 7-4.

[7] In this game one player would hide his or her eyes and then have to guess which of the playmates hands had struck them. The French borders picture two versions of this game; on the left a player turns away and prepares to be struck on their behind, while on the right a player has his eyes covered and holds forth his hands to be struck. For more on this game, see Iona Opie and Peter Opie, *Children's Games in Street and Playground* (Oxford, 1969), 221.

[8] Douce 135 is usually dated to the first half of the sixteenth century. See Henisch, colour plate 7-14, Pächt and Alexander, no. 851, and J. F. Willard, "Occupations of the Months in Medieval Calendars", *The Bodleian Quarterly Record* 74 (1932), 35.

[9] Roger Wieck, *Painted Prayers: The Book of Hours in Medieval and Renaissance Art* (New York, 1997), 33.

small, hand-illuminated manuscript now in Paris, it was through his designs for printed Parisian Books of Hours such as PML 572 that the Master of Anne de Bretagne became renowned.[10] Unlike other manuscript illuminators, the Master of Anne de Bretagne embraced the technology of printing and realised its commercial potential. Particularly striking are his complex border designs, which were traditionally the reserve of luxury manuscripts. An example of this is that a number of the playful scenes which appear in PML 572 and Douce 135 correlate with imagery found in the *Hours of Adélaïde de Savoie*. In the February pages of Douce 135 and PML 572 a group of men and women cook Shrove Tuesday pancakes, an activity which is pictured in the upper border of the February miniature in the *Hours of Adélaïde de Savoie*. In March, all three books illustrate Ash Wednesday with a group of worshippers receiving ashes; in April a group of pilgrims are shown approaching a church, and in all three the month of December features a snowball fight.

It seems likely that these overlaps with the *Hours of Adélaïde de Savoie* were coincidental rather than deliberate. The scenes which depict the same subjects are staged in completely different ways, and in many of the months the subjects differ, with the *Hours of Adélaïde de Savoie* favouring traditional harvest labours rather than games in the months of June, July, August and September. The relationship between PML 572 and Douce 135, however, appears to have been much more direct. Despite this, several differences between the two books suggest that they were produced for different audiences. These include the use of colour in the hand-illuminated scenes versus the monochrome of the printed *Horae* and

---

[10] The designs that the Master of Anne de Bretagne produced for the publisher Simon Vostre in the years 1495-1498, which include the Pierpont Morgan *Horae*, represent some of his best and most widely disseminated works. The designs exerted a wide influence on the decorative arts throughout France, and his style dominated Parisian book decoration into the first decade of the sixteenth century. Because of his varied output, the artist is identified by a variety of different names. I use "Master of Anne de Bretagne" (after the *Très Petites Heures d'Anne de Bretagne*, Paris, BNF, MS nouv. acq. lat. 3120) as this refers to his work as a miniaturist on. For more on this artist, see François Avril and Nicole Reynaud, *Les manuscrits à peintures en France 1440-1520* (Paris, 1993), 265-270 and Ina Nettekoven, *Der Meister der Apokalypsenrose der Saint Chapelle und die Pariser Buchkunst um 1500* (Turnhout, 1994).

variations in the backgrounds and in the number of figures depicted.[11] With fewer figures in each scene the Douce manuscript borders are much clearer and more immediately legible than the crowded, monochrome borders of PML 572. The backdrops to the hand-illuminated borders include verdant countryside, cobbled courtyards and castle walls, all of which display a convincing recession of space. By contrast, in the printed *Horae* most of the games are set against a townscape, where the roofs of the houses make the scenes seem flat and frieze-like. This gives the printed scenes an urban feel and suggests that, like the *Hours of Adélaïde de Savoie,* the original owner of PML 572 might have been bourgeois. By contrast, the sections of castle wall evident in the background of scenes in Douce 135 indicate that the original owner of this manuscript might have belonged to the aristocracy.[12] The different media of the two books support these conclusions. The introduction of printing late in the fifteenth century meant that for a period the production of printed books and illuminated manuscripts overlapped. In order to distinguish themselves from the new print technologies, traditional manuscripts became increasingly lavish. While printed Books of Hours appealed to a wide market, hand-decorated manuscripts were only affordable for the elite. The relationship between Douce 135 and PML 572 thus demonstrates an interesting anomaly, with a luxurious, hand-illuminated manuscript being produced after iconography found in a printed book.

Calendar Games in Ghent-Bruges Manuscripts

The concept of a playful calendar was to become more firmly established within a number of Ghent-Bruges manuscripts. Emerging during the first half of the sixteenth century from the artistic centers of Ghent and Bruges, these beautiful books represented the final flourishing of manuscript illumination. Although their precise authorship is much disputed, the manuscripts can be broadly attributed to the workshops of Gerard Horenbout in Ghent and Simon Bening in Bruges.[13] It is clear that

[11] In the month of February, for example, there are eight figures making pancakes in PML 572 and six in Douce 135, just as in September (illustrated) eight figures play games of *frappé main* in PML 572 and six play in Douce 135.

[12] These include the games of golf (f.2r), blindman's buff (f.5v), "piggy in the middle" (f.7r) and the snowball fight (f.7v).

13 On the Ghent-Bruges school see Kren and McKendrick, parts 4 and 5, Maurits Smeyers and Jan Van der Stock, *Flemish Illuminated Manuscripts 1475-1550*

the artists who worked on the manuscripts were highly skilled. The life-like miniatures resemble tiny panel paintings, with attention paid to realistically rendering the finer details of everyday objects, atmospheric effects, facial expressions and skin tone. The borders of the Ghent-Bruges manuscripts represent important sites for further elaboration. In at least seven manuscripts seasonal play is systematized into a detailed decorative program, with borders featuring a "game of the month" complementing and expanding upon the labours of the months pictured in the main miniatures.[14] The seven extant manuscripts which feature game-calendars include three related manuscripts from the first decade of the sixteenth century which have twelve-page calendars; the *Mayer Van den Bergh Breviary* (Antwerp, Museum Mayer van den Bergh Cod.946, c.1500), the *Eleanor of Portugal Breviary* (New York, Pierpont Morgan Library MS M.52, c. 1510) and the *Spinola Hours* (Los Angeles, Getty Institute MS Ludwig IX.18, c.1515).[15] Two more manuscripts feature seasonal game borders across twenty-four page calendars; the *Lisbon Prayerbook* (Lisbon, Museu Nacional de Arte Antiga MS 13 Illum., c. 1530) and the Munich *Spielbuch* (Munich, Bayerische Staatsbibliothek CLM 28346, c.1500).[16] A sixth manuscript known as the *Hours of Conde-Duque*

(Ghent, 1996), 9-47, and Georges Doegar, *Flemish Miniature Painting in the Fifteenth and Sixteenth Centuries* (Amsterdam, 1987), 13-19.

[14] The phrase "game of the month" was coined in a recent article on the Ghent-Bruges calendar games, see A. Willemsen, "The Game of the Month: Playful Calendars in Ghent-Bruges Books of Hours", in *Manuscripts in Transition: Recycling Manuscripts, Texts and Images*, ed. Bridgette Dekeyzer and Jan Van der Stock (Leuven, 2005), 419-430. The first scholar to pay attention to these manuscript game borders and link them to Bruegel's painting *Children's Games* was Sandra Hindman, see S. Hindman, "Pieter Bruegel's *Children's Games*, Folly and Chance", *The Art Bulletin* 63 (1981), 447-475.

[15] All three of these manuscripts feature twelve-page calendars with depictions of seasonal games executed in gold, silver, bronze, white or coloured *camaïeu* in the upper borders. All three also share the same mistake (a transposition of the zodiacal signs of Scorpio and Libra) and this has been seen as evidence that they originated in the same workshop. Willemsen, "The Game of the Month", 423.

[16] The *Lisbon Prayerbook* is sometimes referred to as the Hours of D. Fernando or of D. Catarina of Portugal, but as the identity of its original owner remains uncertain I refer to it simply as the *Lisbon Prayerbook.* Its calendar is incomplete—two folios, one with the calendar text for June and the labour illustration for July and one with the calendar text for November and labour for December were cut from the manuscript at some point. The June/July folio has been replaced with one showing the appropriate calendar text, with a decorated

Olivares (Valencia, Real Colegio de Corpus Christi, c. 1505) clearly shares some of the same iconography as the other manuscripts but differs somewhat in style.[17] Finally, the British Library's *Golf Book* (London, British Library Additional 24098, c. 1540) contains what is probably the latest and certainly the most elaborate game calendar (fig. 1).[18]

While the main calendar miniatures in the Ghent-Bruges manuscripts often follow the example of the *Très riches heures* in depicting the leisure activities of the aristocracy (hunting, hawking, courting and music-making) in the borders the pastimes of the peasants take precedence. We can imagine that the lively depictions of popular games and pastimes found here provided anecdotal interest as well as pleasure to the elite viewer. As with the labours, the types of games played throughout the year were dictated in part by the weather conditions—hence the months of December, January and February are typically decorated with games involving snow and ice, while in July and August communal outdoor tournaments of archery and water-jousting dominate. In other months, the

---

border of *putti* at play. The verso has been left blank. For more on this manuscript see Willemsen, "The Game of the Month", 421; Kren and McKendrick, 8; J. Black, "Museu Nacional de Arte Antigua", *Descriptive Inventories of Manuscripts Microfilmed for the Hill Monastic Library: Portuguese Libraries*, vol. 3 (Minnesota, 1990), 518-519. The *Spielbuch* was christened by Willemsen; see Annemarieke Willemsen, *Kinder delijt: Middeleeuws speelgoed in de Nederlanden* (Nijmegen, 1998), 215.

[17] See Felipe M. Garin Ortiz de Taranco, *Libro de Horas del Conde-Duque de Olivares: Estudio del códice brujense del Real Colegio de Corpus Christi, en Valencia, y de la ilustracíon europea de su tiempo* (Valencia, 1951). This manuscript demonstrates the greatest deviation from the Ghent-Bruges standard. The large, squarer shape of the borders means that the games are not spread out in a narrow strip but instead are given even greater emphasis than the labours of the months. Parallels between the *Hours of Conde-Duque Olivares* and the *Grimani Breviary* have been made, and certainly the figures are shown in a similar amount of detail; in both cases, for example, facial expressions can be easily read. Yet the figures in the *Olivares Hours* are stiffer than those of the *Grimani Breviary*, their poses robotic rather than naturalistic, so that ultimately this manuscript seems like sincere and innovative imitation of the Bening-Horenbout style by another hand rather than anything that would have come from the workshops.

[18] Named after the depiction of a game of *kolf* (an early form of golf) in the September border, the *Golf Book* consists of the calendar pages, a few miniatures, and some pages with historiated initials, all cut from what would originally have been a lavish Book of Hours. The sheets were rebound and refoliated in 1928. See Kren and McKendrick, no. 155.

border games bear a more direct relationship to the labour of the month. In the month of October in the Mayer Van den Bergh Breviary we see cattle being sold and slaughtered in the lower border while in the upper border figures play games with animal parts: skittles with knucklebones on the left and a fight with inflated bladders on the right (fig. 4). Although there is a strong degree of uniformity in the order in which the seven manuscripts present their seasonal games, there is considerable scope for variation within the staging of the individual scenes. Many of the March game borders picture games with whip-tops and rattles, but we can observe considerable differences in the composition and style of each border (figs. 5, 6 and 7). This suggests that the game borders were not executed using workshop patterns. Instead, it seems that the artists were given free rein to depict the traditional activities in their own ways. This raises the question of how the seasonal game calendar was established and understood. Do the game-borders accurately reflect patterns of leisure time, or are they an independent artistic convention with little basis in the realities of late-medieval life?

We can explore this further by looking at the games which decorate the month of March, considering how these might relate to broader concepts of belief and time, and also to depictions of the same games in Bruegel's paintings. The games which feature in the March borders typically involved whip-tops or rattles, with all of the manuscripts picturing at least one and more often both of these games.[19] The purpose of a calendar in a Book of Hours was to highlight Holy Days, and as such it is perhaps unsurprising that some of the games pictured in the borders relate to religious rituals. Both the activities of waving a rattle and whipping a top can be connected to the liturgical period of Lent. Further evidence of this is that games with whip-tops and rattles appear in Bruegel's *Battle Between Carnival and Lent* (Vienna, Kunsthistoriches Museum), a

[19] The game occurring in the March borders of the *Mayer van den Bergh Breviary* and the Munich *Spielbuch,* and in the February border in the *Hours of Conde-Duque de Olivares* is often identified as whip-top but might be more accurately labelled "blind shoe". Depictions of this game, in which a central blindfolded figure holds a shoe suspended on a rod above his playmates' heads, appear in a number of other sixteenth-century sources, including Bruegel's panel *Children's Games*, a Southern Netherlandish *millefleurs* tapestry in the Metropolitan Museum of Art, an ivory diptych sundial made in Nuremberg and a French Book of Hours in the Bodleian Library (Oxford, Douce MS 276, f. 59v).

painting which catalogues the customs and rituals of these periods.[20] We know that Carnival activities would gradually build throughout the winter months of January and February and culminate on Shrove Tuesday. The following day, Ash Wednesday, marked the beginning of Lent, when all excesses were banned for forty days until Easter, which usually fell in the middle of March. Bruegel's painting stages the climactic moment when Shrove Tuesday and Ash Wednesday meet in a mock-joust. Although evidence suggests that jousts among costumed revellers did occur at Carnival time, Bruegel's painting is in many ways full of artifice. The varied events of the periods of Carnival and Lent are effectively compressed into a single confrontation, with the joust serving as a device around which two antithetical states are detailed. Beyond the fat figure of King Carnival, the left side of the panel is filled with villagers who are eating, drinking, making traditional food, dancing and performing plays. Behind Queen Lent, who is personified by a mean-faced and emaciated nun, we see pious villagers spring cleaning, visiting the church and giving alms to beggars. The onset of spring on the right side of the panel is also suggested by the trees, which are beginning to sprout leaves. The whip-tops and rattles which featured in the March calendar borders are present here—almost all of Lent's followers hold rattles, and beside the church a group of children play with whip-tops.

The links between rattles and the period of Lent are fairly clear. Church bells were silenced from Maundy Thursday until Saturday morning mass, and during this period the bells would be replaced by rattles, which were often played by children.[21] This was obviously undertaken enthusiastically—in 1520 the city council of York had to take measures against boys whose Easter rattles disturbed the peace.[22] However, visual evidence suggests that this activity was not solely restricted to children. In Bruegel's *Carnival and Lent* and in the manuscript borders the rattles are clutched by both large and small figures alike, suggesting a communal cacophony of noise. Slight differences can be observed in the ways that the rattles are held. Most of the figures in the manuscripts hold their rattles horizontally, with a hand on either side, whereas the rattles in Bruegel's painting are held vertically and from

[20] For detailed reproductions of all of the Bruegel paintings in Vienna see W. Seipel, *Pieter Bruegel the Elder at the Kunsthistoriches Museum in Vienna* (Vienna, 1998).
[21] M. Rozler, "Rattles on Good Friday and Maundy Thursday in Austria", *Folklore*, 47 (1936), 393-394.
[22] Nicholas Orme, *Medieval Children* (New Haven, 2001), 186

underneath, probably by some kind of handle. In both cases the group of figures advance in a line, although the processional nature is more explicit in Bruegel's work as the players are clearly following Lent's cart. Their link to Lent is made clear by a little girl who leads the group. She grasps her rattle in both hands and looks up at the figure of Lent, a sooty cross painted on her forehead. The rattles most commonly shown are wooden "clappers" with rocking arms that would resound on the wooden base when shaken. An illustration from a sixteenth-century French manuscript in the Bodleian library, Douce MS 276, shows similar rattles, this time with double rocking arms (fig. 8). This image also details a second type of rattle in evidence in the Ghent-Bruges borders. This appears to be a type of clacker, the ridged end of which would be whirled around to create a noise.[23] Again, the liturgical aspect of the activity is underlined by the fact that the figures in the French manuscript are entering a church.

Whip-tops also have a long association with Shrovetide and the month of March, although the reasons behind this are less clear. It has been suggested that the public holidays associated with the religious festivals of Shrove Tuesday and Good Friday might have stimulated game playing, although this does not explain the predominance of the whip-top in particular. In the early sixteenth century Alexander Barclay described the seasonal activities of children in rhymed verse, writing:

> Each time and season hath his delights and joys,
> Look in the streets behold the little boys,
> How in fruit season for joy they sing and hop,
> In lent is each one full busy with his top.[24]

Although, as Barclay's verse suggests, boys traditionally played with whip-tops, in Bruegel's painting the boys are joined by girls, and their game is placed beside the church on the Lenten side of the panel. One reason for the occurrence of this game in the month of March seems to have been the weather. In the spring months children would have been encouraged to play by the improved weather, but it would still have been fairly cold. In these conditions the activity of whipping a top was a practical way to keep warm; in 1600 Samuel Rowlands recommended

---

[23] A. Willemsen, "Medieval Children's Toys in the Netherlands: Production, Sale and Trade", in *Material Culture in Medieval Europe* (Zellik, 1997), 405-411.

[24] Developed from the sixth *Eclogues* of the Italian humanist Mantuan. See Beatrice White, ed., *The Eclogues of Alexander Barclay* (London, 1928), 5th Eclogue, lines 87-90.

"toppe and scourge to make him hot" and the fact that whip-tops continued to be played with in springtime into the twentieth century confirms this wisdom.[25] It is also possible that, like the rattle, the whip-top had religious associations which made it pertinent to the Easter period. Some scholars have suggested that the activity of whipping a top might have been encouraged during Lent because of the way in which it echoed the flagellation of Christ.[26] According to the Bible, Christ was flagellated on Maundy Thursday, prior to his crucifixion on Good Friday. The flogging of Christ by his tormentors was a popular subject in art at this time; in representations Christ is pictured tied to a pillar and attacked by soldiers armed with whips and scourges. The religious dimension to the game of whip-top would seem to be supported by a number of manuscripts which feature images of children whipping-tops in front of exterior views of churches.[27] In Bruegel's *Children's Games* boys whipping tops are also placed beneath the ecclesiastical arches of the loggia attached to the central building.

However, the whip-top was by no means tied to religious narratives. As well as appearing in *Carnival and Lent* and *Children's Games*, whip-tops feature in several of Bruegel's winter scenes, where the toy functions, like the skate and the sled, as a generic attribute of winter play. One such example is the children playing in the foreground of *The Census at*

---

[25] Samuel Rowlands, The *Letting of Humours Blood in the Head Vaine* (1600), cited in Iona Opie and Peter Opie, *Children's Games with Things* (Oxford, 1997), 309.

[26] Edward Snow, *Inside Bruegel: The Play of Images in Children's Games* (New York, 1997), 115-116; Orme, 186-187.

[27] Willemsen discusses how this game transcends the calendar-borders and is illustrated elsewhere in a number of manuscripts. Games with whip-tops occur beneath a depiction of "St Anthony of Padova" in the *Rothschild Prayer Book* (Present location unknown, formerly in Vienna, Österreichische Nationalbibliothek, Ms. S. n. 2844), beneath an image of "St John on Patmos" in the *Stockholm-Kassel Book of Hours* (Stockholm, Kunglige Biblioteket, Ms. A227), beneath the "Visitation" in a manuscript in Lisbon (Museu Nacional de Arte Antigua, Ms. 13 Illum., f. 32v), and beneath the "Vera Icon" in the *Imhof Prayer Book* (auctioned at Sotheby's in London on 21 June 1988, lot 107). The motif appears twice in the *Book of Hours of James IV of Scotland* (Vienna, Österreichische Nationalbibliothek, Ms. 1897, f. 18 and f. 190). Willemsen proposes that one reason for this may be because it was an attractive way to fill a bottom and side margin with one continuous scene; with a beautiful architectural façade stretching up the outer margin and the anecdotal game detail adding interest to the lower margin. See Willemsen, "The Game of the Month", 426-427.

Bethlehem (Brussels, Musées Royaux des Beaux-Arts). Although this painting has a religious narrative—Mary and Joseph arrive at the census in the foreground—it seems unlikely that the children whipping tops beside them are included as a coded reference to Christ's fate. Instead, the playing children set up a counter-narrative to the religious group, their gleeful games providing both anecdotal interest and authenticity to the contemporary Flemish village that is Bruegel's Bethlehem. We can imagine that a frozen river such as the one on which the children play would have provided an ideally smooth surface on which to turn a top, and, as we have already seen, the physicality of whipping a top was also a simple way to stay warm in the winter months. The whip-top, like the hobby-horse and windmill, was a toy which had long functioned as a generic symbol of childhood or childish behaviour. The two components of the wooden top and the handled whip made this game easily identifiable, whether as a simple line-drawn drollery or as a distant silhouette in a wintry landscape. In the Tudor morality play *Mundus et Infans*, the first game that the child, Infans, plays is whip-top.[28] It is likely that his playful speech, in which he describes how he uses his scourge-stick to hit his fellow playmates, would have been accompanied by lively actions on the part of the actor. Bruegel's preference for the whip-top over the spinning top can perhaps also be related to the theatricality inherent in the action of whipping a top. Other paintings to feature the whip-top are the *Winter Landscape with Bird Trap* (Brussels, Musées Royaux des Beaux-Arts) and the *Hunters in the Snow* (Vienna, Kunsthistoriches Museum), where children can be found whipping tops on the frozen river beside adults engaged in a game which resembles curling.

Whip-tops and rattles are not the only seasonal games to appear in Bruegel's paintings. A series of paintings produced in the 1560's which depict the months of the year are closely connected to the iconography of the Ghent-Bruges calendars. Commissioned by the Antwerp tax collector Niclaes Jongelinck and hung in his suburban villa *Ter Berke*, the series of panels known as the *Months* (five of which are now extant) mimic manuscript pages in their atmospheric presentation of the changing seasons and tasks.[29] As in Ghent-Bruges calendar iconography, it is the landscape and labours which dominate. But looking closely we find that in

[28] G. A. Lester, *Three Late Medieval Morality Plays: Mankind, Everyman, Mundus et Infans* (London, 2002), 115.

[29] I. Buchanan, "The Collection of Niclaes Jongelinck: II. The 'Months' by Pieter Bruegel the Elder", *The Burlington Magazine* 132 (1990), 541-550.

four of the five panels the labours are complemented by tiny depictions of seasonal games and festivities. These most often appear in the deep-space of the background landscape. In the *Hunters in the Snow* the figures skating, sledging and whipping tops on the frozen river are barely more than silhouettes. In *The Harvesters* (New York, Metropolitan Museum) the playing figures are fainter still, but we can still make out the details of their games—swimming and "cock-throwing", both activities common to the summer manuscript borders—in the green fields to the left of the panel.[30] A third panel from the *Months* series, *Haymaking* (Prague, National Gallery), includes in the background a large group of archers shooting upwards at a popinjay (a bird mounted on a mast), an activity which is also depicted in the May borders of the *Olivares Hours*, the Munich *Spielbuch* and the *Golf Book*. A fourth panel, *The Gloomy Day* (Vienna, Kunsthistoriches Museum), adopts a composition which mimics the layout of the Ghent-Bruges manuscript pages even more closely. Here the festive activities associated with Carnival time appear to bracket the central labours of pruning trees and bundling wood, with figures dancing and playing the bagpipes in the lower left corner of the panel and a family

[30] The essence of cock-throwing involved throwing a stick at a bird to stun it. Although outlawed in the nineteenth century, evidence from manuscripts suggests that this game was played as early as the thirteenth century. A single figure is playing the game in a French manuscript which dates from the late thirteenth century in the Bibliothèque Nationale (Robert de Borron, *L'Histoire du grail*, MS. Fr. 95), two apes take aim at a bird in a fourteenth-century manuscript from the Walters Gallery (MS. 82), and two men take aim in a manuscript in the Brussels Bibliothèque Royale (Bruno Latini's *Trésor des* Sciences, MS 10228). See Lillian Randall, *Images in the Margins of Gothic Manuscripts* (Berkeley, 1966). The game is clearly being played in the August border of the *Golf Book*, and is possibly depicted in some of the other game-borders in September (see Buchanan, "The Collection of Niclaes Jongelinck", 546). On the rules and popularity of the game, see Joseph Strutt, *The Sports and Pastimes of the People of England* (London, 1876), book 3, chapter 7, no. 21 and Robert Malcolmson, *Popular Recreations in English Society 1700-1850* (Cambridge, 1973), 119-122. The group cock-throwing in this image includes men, women and children, but it is the men and women who are armed with sticks and ready to take aim while the children seem to be merely observing the game; one small figure runs excitedly between the throwers and the bird, and others huddle around the players. Children play a similar role in a sixteenth-century drawing depicting cock-throwing attributed to the Master of the Small Landscapes. For an image see Nadine Orenstein, ed., *Pieter Bruegel the Elder: Drawings and Prints* (New York, 2001), no. 139, *Landscape with Figures throwing Clubs at a Goose.*

eating waffles in the lower right. It is likely that the manner in which the *Months* panels recall manuscript calendars was entirely deliberate. Ghent-Bruges manuscripts were incredibly expensive and prestigious objects, and their ownership was largely limited to members of the great courts of Europe, most notably the Burgundian and Habsburg dynasties. It has been suggested that in commissioning the series from Bruegel the socially ambitious Jongelinck was quite consciously emulating iconography more typically associated with aristocratic culture.[31] In both cases, we can imagine that the images of popular games would have entranced and entertained sophisticated viewers, adding an extra dimension to the more established iconography of the seasonal labours.

New ordering principles: Bruegel's *Children's Games*

We have seen how popular games feature in the background of a number of Bruegel's paintings, but in *Children's Games* they take centre stage. With over two hundred playing figures depicted on a panel a metre high by a metre and a half wide, *Children's Games* is the sixteenth century's most visually ambitious representation of games. The games in the Ghent-Bruges manuscripts provide an obvious correlation with this work: many of the same games are shown, and moreover they share a life-like visual style which is distinguished by close attention to the details of clothing, facial expressions and play-objects. Manuscript game borders supply visual precedents for the vast majority of toys which are found in Bruegel's panorama, from manufactured items (hoops, whip-tops, rattles, hobbyhorses, windmills, marbles, bowls, stilts and pigs' bladders) to found objects (bones, bricks, rushes, pips and stones). In several instances manuscript depictions elaborate upon how some of Bruegel's more unusual games might have been played, including, for example, the game of "blind-shoe".[32] Furthermore, the way in which the illustrious Books of Hours and Breviaries serialized children's play is suggestive of a context

[31] For more on these panels and the original context in which they would have been displayed—see C. Goldstein, "Artifacts of Domestic Life: Bruegel's Paintings in the Flemish Home", *Nederlands Kunsthistorisch Jaarboek* 51 (2000), 174-193; Buchanan, "The Collection of Niclaes Jongelinck"; K. Moxey, "Pieter Bruegel and Popular Culture", in David Freedberg, ed., *The Prints of Pieter Bruegel the Elder* (Tokyo, 1989), 50.

[32] See note 18.

in which elite audiences enjoyed looking at representations of popular games.

It is equally instructive to note the areas of divergence between Bruegel's panel and the manuscript borders. Absent from *Children's Games* are all of the games in the manuscripts that were played by adults, including shooting at a popinjay and cock-throwing. We know that Bruegel was aware of these games and their place in the seasonal calendar; they appear, as we have seen, in the *Months* panels, and are also depicted in several prints after designs by Bruegel.[33] We can thus see the absence of games of cock-throwing and archery from *Children's Games* as a deliberate exclusion, suggesting that the painting was not ordered by the same principles as the manuscript borders. Further evidence of this can be found in a closer examination of the composition of *Children's Games*. While there are occasions where seasonal games appear side-by-side, this appears to be more by chance than design, for there are as many examples of seasonal games being split up and depicted on opposite sides of the panel.[34] A further paradox can be found in the relationship between the background landscape and the collection of game motifs in *Children's Games*. While the light sky, winding river and trees in full leaf connote summer, February hoops, March whip-tops and rattles and autumnal games with pigs' bladders and knucklebones all feature among the painting's many motifs. Winter games such as skating and snowball fights are the only ones to be excluded, probably because it would have been impossible to integrate them plausibly into the summer landscape. We can thus conclude that in producing *Children's Games* Bruegel drew on manuscript traditions but rejected their rigid ordering principles, producing instead a collection which was both more encyclopedic in scope and more random in arrangement than the manuscript borders. Bruegel's panel also shifts the focus of game-playing away from the seasons and towards the culture of children.

---

[33] Archers aim at a popinjay in the background of the *Kermis of Saint George* (c. 1559) and in the foreground of the *Kermis at Hoboken* (after an original drawing in London dated 1559) a group of adults with longbows shoot at a target while a group of children play a game of span-counter. See Orenstein, no. 79 and no. 80.

[34] Examples where seasonal games appear together include the boy wearing a rush hat and blowing bubbles (both activities associated in the game-borders with the month of June) and the proximity of the boys whipping-tops to the girl with the rattle (both activities associated with March). Seasonal games that are separated include the games with knucklebones and the pigs' bladder and the windmill joust and hobbyhorse.

Further evidence of these points can be found in the kinds of festive traditions which are incorporated into *Children's Games*. Previous scholars have linked festive details in the panel to the summery weather, surmising that the painting is set in midsummer.[35] The bonfire just visible in the receding street on the right of the panel was a traditional staple of the celebrations of St John's Day, which occurred near the midsummer solstice on June 23 and 24. The children seen going from house to house near the bonfire have been interpreted as asking for logs for the fire, another ritual associated with St John's Day festivities.[36] Near to them, a procession of four children carry aloft forked sticks with paper attached—a ritual which corresponds to recorded celebrations for the related Feasts of Saints Peter and Paul on June 29.[37] Finally, the children playing in the river may be a further reference to St John the Baptist, or "Sint jan de wasscher" as he was called in Flanders.[38]

However, despite containing these details, it is important to note that *Children's Games* does not depict either of the summer festivals unequivocally. Most of the festive motifs are positioned in the distant street and are not the focal points of the composition; in fact, many of the figures found here are engaged in activities which appear to be almost deliberately ambiguous. Furthermore, key details traditionally associated with the midsummer festivals are absent from the panel. Accounts of the St. John's Day bonfire often describe the ritual of young men jumping over the fire, a tradition which easily lent itself to visual representation but which is absent from *Children's Games*.[39]Absent also is the dancing,

---

[35] Hindman, *Pieter Bruegel*; Charles de Tolnay, *Pierre Bruegel l'ancien*, 2 vols. (Brussels, 1935), 25-26.

[36] Hindman, *Pieter Bruegel*, 453; Otto von Reinsberg-Düringsfeld, *Calendrier Belge: Fêtes religieuses et civiles, usages croyances et practiques populaires des Belges anciens et modernes*, 2 vols. (Brussels, 1891), vol. I, 417.

[37] Ibid., 439.

[38] Ibid., 420-421.

[39] "On the Vigil of Saint John the Baptist, commonly called Midsummer Eve, it was usual in most country places, and also in towns and cities, for the inhabitants, both old and young, and of both sexes, to meet together, and make merry by the side of a large fire made in the middle of the street, or in some open and convenient place, by way the young men frequently leaped by way of frolic, and also exercised themselves with various sports and pastimes, more especially with running, wrestling and dancing." Strutt, 462. Reinsberg-Düringsfeld states that young and old alike leapt over or danced around the midsummer bonfire to purify and protect themselves from illness. Reinsberg-Düringsfeld, vol. I, 414. Fire-leaping is illustrated in Jacques Stella's *Jeux et plaisirs d'enfance*, a collection of

feasting and garlands of flowers which are traditionally mentioned in accounts of midsummer festivities.

We have seen how in the *Battle Between Carnival and Lent* Bruegel pictured liturgical traditions from different times of the year side-by-side, creating a composite image *about* rather than *of* Carnival and Lent. A similar approach seems to have been employed in *Children's Games*. For although it is the summer festivities that have received the most attention from scholars, *Children's Games* also includes details from a number of winter festivals. Three children in the panel wear the paper crowns which were traditionally associated with the winter festivals of Carnival and Epiphany.[40] These crude crowns were made from lengths of hand-painted or printed paper wrapped around the front of the wearer's head, and were sold door-to-door by children on Twelfth Night (5 January) and Epiphany (6 January).[41] The crowns are worn by children in other works by Bruegel which reference Carnival time, including a small boy at the lower left of *Carnival and Lent* and a small boy at the lower right of the *Gloomy Day* (Vienna, Kunsthistoriches Museum). Although these are often thought to be the earliest depictions of paper crowns, they are predated by a boy wearing a paper crown in Pieter Aertsen's *Peasant Company by the Fireside* (1556, Antwerp, Mayer van den Bergh Museum).[42] Again in this painting the paper crown signifies winter rather than summer festivities.

One of the crown-wearers in *Children's Games* holds another object which does not fit the midsummer theme—a *duivekater* loaf. These

---

52 images of *putti* playing games which were engraved and published with accompanying verses by Stella's niece, Claudine Bouzonnet, in 1657. Beneath the illustration of fire-leaping (no. 11) the verse reads: "To them it seems a wondrous treat, Right in the heart of summer's heat, To leap across a brushwood fire; But if the pleasure of this game, Should last no longer than their flame, It will soon gutter and expire"—see Stanley Appelbaum, *Games and Pastimes of Childhood: An Unabridged Reprint of the 1657 Les jeux et plaisirs d'enfance with 52 illustrations by Jacques Stella* (New York, 1969).

[40] Paper crowns are worn by the child inflating a bladder in the right foreground, by the child holding a loaf of bread and standing on the green ground beyond the beam, and by the right-hand player in couple playing "scissors, paper, stone" to the left of the central group of tug-of-war.

[41] For examples of seventeenth-century printed woodblock Epiphany crowns and depictions of the crowns in paintings see M. Royalton Kisch, "The King's Crown: A Popular Print for Epiphany", *Print Quarterly* 1 (1984), 43-51. See also Reinsberg-Düringsfeld, 24.

[42] Ethan M. Kavaler, *Pieter Bruegel: Parables of Order and Enterprise* (Cambridge, 1999), 116.

currant loaves were baked for a variety of special occasions, and were especially common during the winter holiday season, which began with the Feast of Saint Nicholas on 6 December and ended with Epiphany.[43] A similar child wearing a Twelfth-Night paper crown and carrying a large *duivekater* loaf appears in Sebastian Vranckx's *Winter*, indicating that both of these attributes connote winter festivities.[44] *Duivekater* loaves feature prominently in a number of paintings by Jan Steen, including the *Feast of Saint Nicholas* (Amsterdam, Rijksmuseum), which depicts many of the rituals associated with the midwinter festival of Saint Nicholas.[45] On Saint Nicholas Eve (5 December) children would leave out their shoes. If they had been good, then they would find their shoes filled with sweets and toys; bad children would receive only a dry switch. Several of the children in Steen's painting hold new toys, but on the left of the painting we see the tearful reaction of a boy whose shoe has been filled with twigs.[46] The two baskets hanging from the right-hand window above the loggia in *Children's Games* might be understood in relation to these St. Nicholas Day traditions.[47] Their contents certainly recall those depicted in Steen's painting; the basket on the right is filled with objects, including a pair of shoes, while a dry bundle of twigs protrudes from the basket on the left. Unsurprisingly, the boy at the window is reaching for the full basket.

Midsummer, midwinter and Carnival are not the only festive periods referenced in *Children's Games.* The mock-marriage procession near the centre of the panel recalls a festive tradition common to yet another time of the year—the springtime custom of choosing and crowning a Whitsun queen.[48] Whitsun or Pentecost fell fifty days after Easter Sunday, usually

---

[43] A huge *duivekater* appears on the counter in Job Berckheyde's *Bakery Shop*, c.1680. See Donna R. Barnes and Peter G. Rose, *Matters of Taste: Food and Drink in Seventeenth-Century Dutch Art and Life* (New York, 1984), no. 4.

[44] Belgium, private collection. Reproduced in A. A. Wagenberg-ter Hoeven, "The Celebration of Twelfth Night in Netherlandish Art", *Simiolus*, 22 (1994), fig. 23.

[45] A *duivekater* loaf also appears in Steen's *Feast of Saint Nicholas*, c. 1667-68 (Rotterdam, Museum Boymans-van Beuningen) and in the still-life display of bread in *The Leiden Baker Arend Oostwaert*, c.1658 (Amsterdam, Riksmuseum).

[46] Perry Chapman, Walter T. Kloek and Arthur K. Wheelock, Jr. (ed.), *Jan Steen: Painter and Storyteller* (New Haven, 1996), no. 30.

[47] Hindman identifies these as St. John's Baskets, in which women placed herbs to bring good fortune on St. John's Day, however the contents of the baskets do not resemble herbs. See Hindman, "Pieter Bruegel", 53.

[48] Hills and Hindman argue that the missing elements of May Day flower garlands and a white dress mean that the bridal procession does not belong within this tradition. Hindman relates the procession instead to depictions of adult weddings in

in the month of May. The ritual of selecting a May Queen or Whitsun-bride is depicted in almost all of the Ghent-Bruges borders in the month of May, where processions form around crowned figures holding branches. Whitsun processions were also painted by Pieter Brueghel the Younger and David Vinckboons, and looking at these helps to illuminate how Bruegel's procession relates to the standard iconography.[49] A painting of *The Whitsun Bride* by Pieter Bruegel the Younger (New York, Metropolitan Museum) shows a procession of children led by a little girl with long hair and a crown. Similarities between this painting and the wedding procession in *Children's Games* include the bride's loose hair and crown, a tradition in the Netherlands dating back to at least the fifteenth century.[50] Other similarities include the demeanors of the participants, from the attentive adults who chaperone each of the processions to the solemn children, many of whom have flipped their skirts over their heads to form make-shift headdresses. There are also some differences between the bridal procession in the New York painting and the motif in *Children's Games*. In later depictions of Whitsun-brides, the Whitsun flowers (*pinksterbloem*) play a large role, being incorporated into headdresses and bouquets. By contrast, the flowers in Bruegel's image are limited to two baskets which are carried by attendants at the front of the procession. Absent from Bruegel's depiction are the fiddle player and drummer who lead the party in the New York painting. Another difference is the dress worn by Bruegel's child-bride, which is black, the colour of Flemish wedding dresses, rather than the traditional Whitsun white.[51] Despite these differences, the frequency of May Queen processions in the manuscript borders and the depictions of Whitsun-brides by later artists suggest that Bruegel's motif could be interpreted in relation to festive, springtime traditions, rather than the moralistic mimicry of adult actions to which it is usually ascribed.

Bruegel's oeuvre. Jeanette Hills, *Das Kinderspelbild von Pieter Bruegel d. Ä., 1560; eine volkskundliche Untersuchung, Veröffentlichungen des Österreichischen Museums für Volkskunde, x* (Vienna, 1957), 29-30 and Hindman, 452.

[49] For these images see Klaus Ertz, *Pieter Brueghel der Jüngere: Die Gemälde mit kritischem Oeuvrekatalog*, 2 vols (Luca, 1999-2002), cat. No. 1042-1053.

[50] W. S. Gibson, "Some Notes on Pieter Bruegel the Elder's *Peasant Wedding Feast*", *The Art Quarterly*, 28 (1965), 194-197 and M. M. Salinger "*The Whitsun-Bride* by Pieter Brueghel the Younger", *The Metropolitan Museum of Art Bulletin*, 34 (1939), 88-90.

[51] Hindman, "Pieter Bruegel", 452, n. 31.

Although we do not know who commissioned *Children's Games*, it seems likely that, like Jongelinck, they might have enjoyed the associations that Bruegel's painting makes with the high status iconography of the Ghent-Bruges manuscripts. However, *Children's Games* is not slavish in its references to manuscript precursors. Referencing the luxurious Ghent-Bruges manuscript iconography, the painting also presents significant innovations. Spanning the calendar from spring to Epiphany, the painting includes a vast range of seasonal games and festive traditions. The only thing which all of these games and festive traditions have in common is their specific relationship to the culture of children. The absence of adult games such as cock-throwing and archery which were common to the manuscript calendars provides further evidence that in *Children's Games* Bruegel sought to picture only those games and traditions which related specifically to childhood. Bruegel's rejection of the temporal organizing strategies employed when depicting popular pastimes in the manuscripts, and his focus solely on the games of children, suggests that by the middle of the sixteenth century the seasonal aspect of game-playing was waning; paving the way for a new, humanistic approach to childhood games which embraced the positive force of play.

Illustrations

Figure 1, January. British Library, London, *Golf Book*, © British Library Board, Additional 24098, fol. 18v.

Figure 2, September. The Pierpont Morgan Library, New York, PML 572, Horae: ad usum Romanum (Rome), fol. 6v.

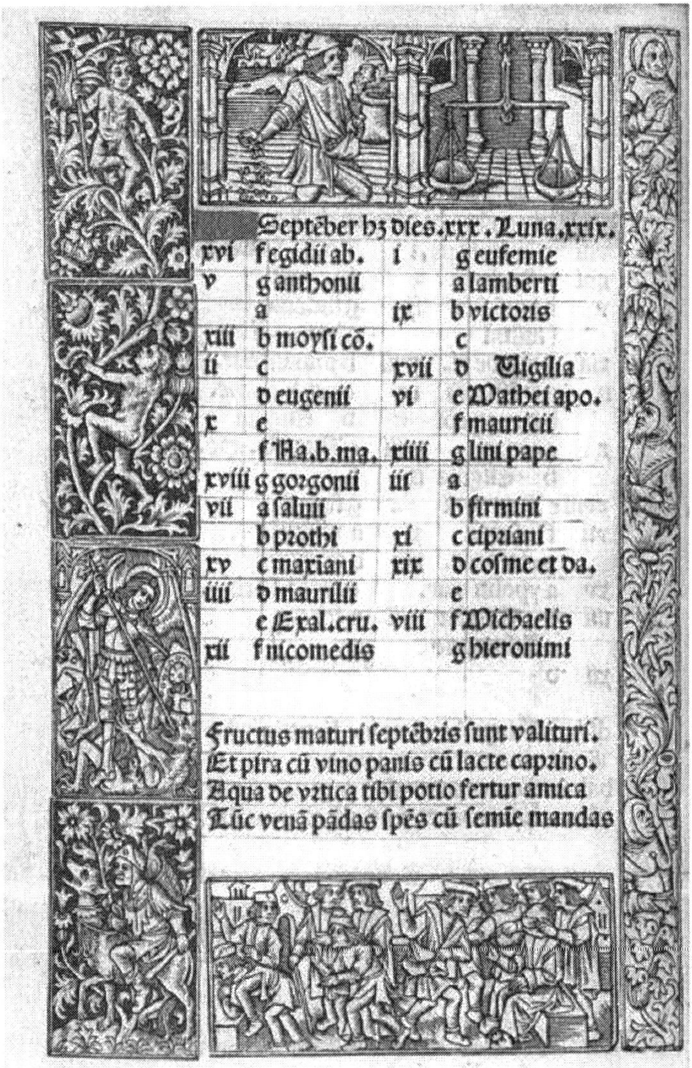

Figure 3, September. The Bodleian Library, Oxford, Douce MS. 135, fol. 6r.

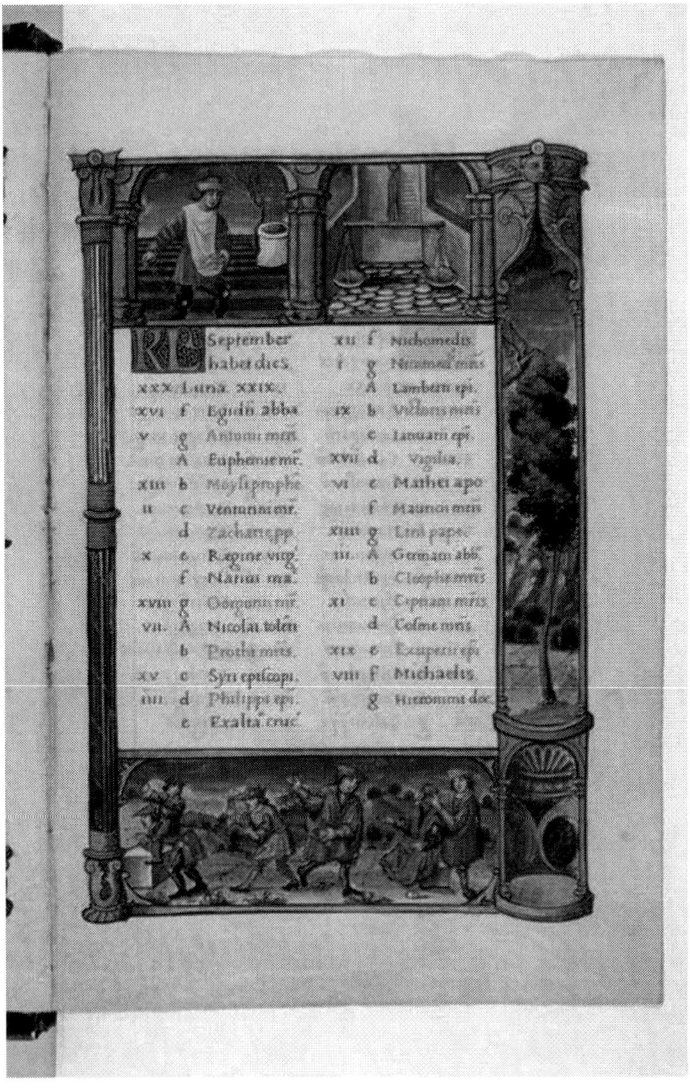

Figure 4, October. Mayer van den Bergh Museum, Antwerp, *Mayer van den Bergh Breviary*, Codex 946, fol. 6r.

Figure 5, Detail from month of March. Museum Mayer van den Bergh, Antwerp, *Mayer van den Bergh Breviary*, Codex 946, fol. 2v.

Figure 6, Detail from month of March. The J. Paul Getty Museum, Los Angeles, Workshop of the Master of James IV of Scotland, *Spinola Hours,* MS 83.ML.114, fol. 2v. About 1510-1520, tempera colours, gold, and ink on parchment bound between pasteboard covered with red morocco, leaf 23.2 x 16.7cm.

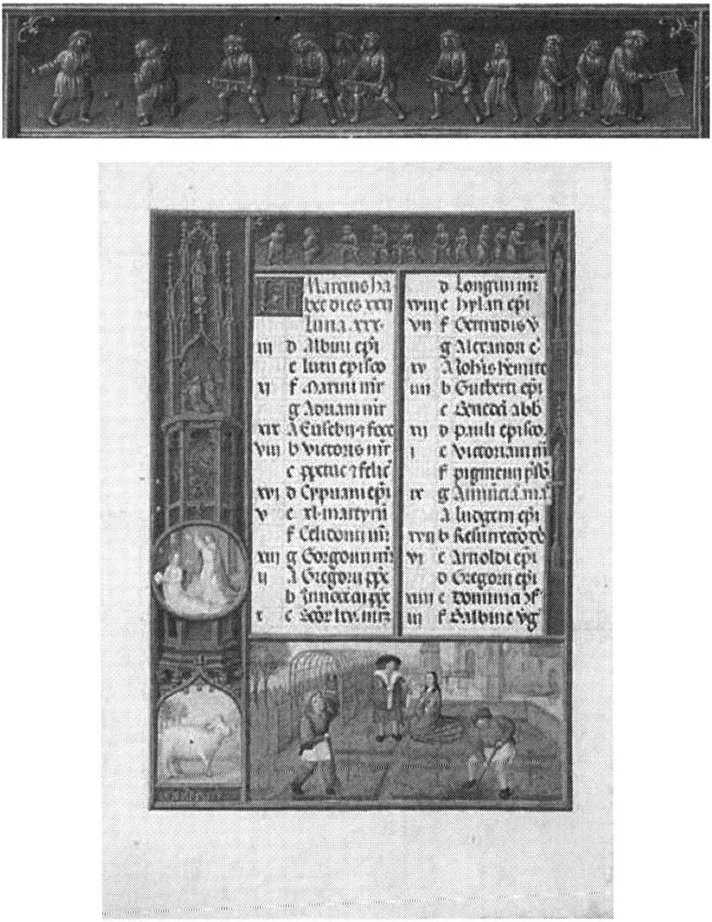

Figure 7, Detail from month of March. British Library, London, *Golf Book*, © British Library Board, Additional 24098, fol. 20v.

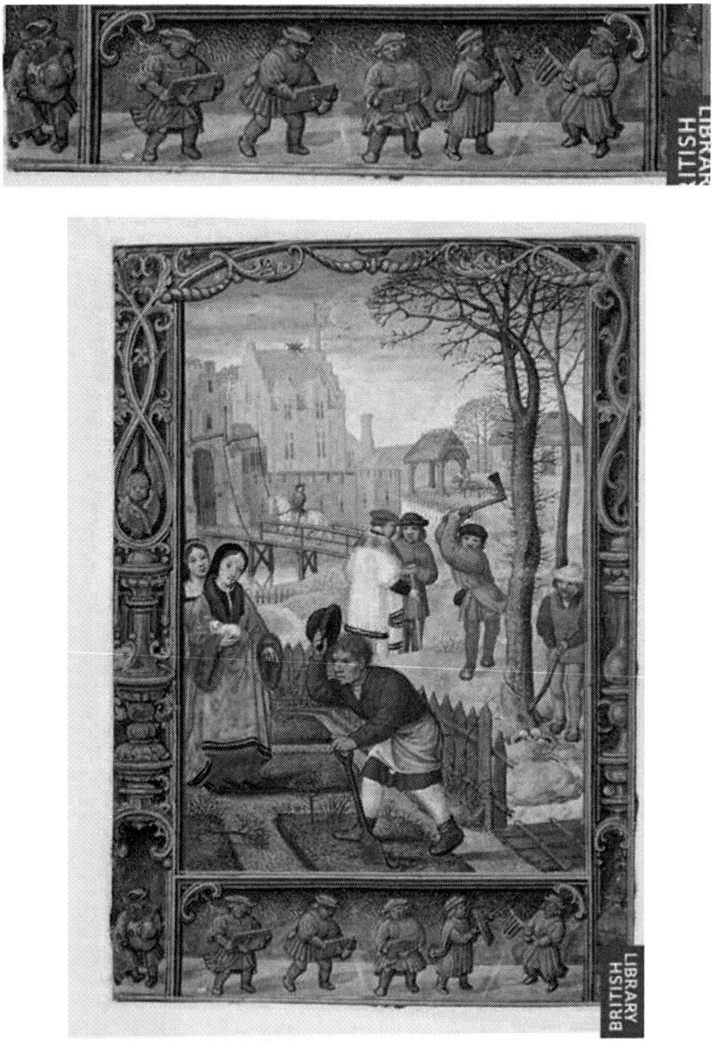

Figure 8, Figures with rattles entering a church. Bodleian Library, Oxford, Douce MS 276, Oxford, fol. 83r.

Bibliography

Alexander, J. J. G., "*Labeur* and *Paresse*: Ideological Representations of Medieval Peasant Labor", *Art Bulletin* 72 (1990), 443-452.

Appelbaum, Stanley, *Games and Pastimes of Childhood: An unabridged reprint of the 1657 Les jeux et plaisirs d'enfance with 52 illustrations by Jacques Stella* (New York, 1969).

Avril, François and Nicole Reynaud, *Les manuscrits à peintures en France 1440-1520* (Paris, 1993).

Barnes, Donna R. and Peter G. Rose, *Matters of Taste: Food and Drink in Seventeenth-Century Dutch Art and Life* (New York, 1984).

Black, Jonathan and Thomas L. Amos, *Descriptive Inventories of manuscripts Microfilmed for the Hill Monastic Library: Portuguese Libraries, Vol. 3, Part 3, Mss 302-456 and Museu Nacional de Arte Antigua, Lisboa* (Collegeville, 1990).

Bouissounouse, Janine, *Jeux et travaux d'apres un livre d'heures du XV siecle* (1925; 2nd edn., Geneva, 1977).

Buchanan, I., "The Collection of Niclaes Jongelinck: II. The 'Months' by Pieter Bruegel the Elder", *The Burlington Magazine* 132 (1990), 541-550.

Cazelles, Raymond and Johannes Rathofer, *Illuminations of Heaven and Earth: The Glories of the Très Riches Heures du Duc de Berry* (New York, 1988).

Chapman, Perry, Walter T. Kloek and Arthur K. Wheelock, Jr., eds, *Jan Steen: Painter and Storyteller* (New Haven, 1996).

Doegar, Georges, *Flemish Miniature Painting in the Fifteenth and Sixteenth Centuries* (Amsterdam, 1987).

Ertz, Klaus, *Pieter Brueghel der Jüngere: Die Gemälde mit kritischem Oeuvrekatalog*, 2 vols (Luca, 1999-2002).

Garin Ortiz de Taranco, Felipe M., *Libro de Horas del Conde-Duque de Olivares: Estudio del códice brujense del Real Colegio de Corpus Christi, en Valencia, y de la ilustracíon europea de su tiempo* (Valencia, 1951).

Gibson, Walter S., "Some Notes on Pieter Bruegel the Elder's *Peasant Wedding Feast*", *The Art Quarterly* 28 (1965), 194-208.

Goldstein, C., "Artifacts of Domestic Life: Bruegel's Paintings in the Flemish Home", *Nederlands Kunsthistorisch Jaarboek* 51 (2000), 174-193.

Henisch, Bridget, *The Medieval Calendar Year* (Pennsylvania, 1999).

Hills, Jeanette, *Das Kinderspelbild von Pieter Bruegel d. Ä., 1560; eine volkskundliche Untersuchung, Veröffentlichungen des Österreichischen Museums für Volkskunde, x* (Vienna, 1957).

Hindman, S., "Pieter Bruegel's *Children's Games*, Folly and Chance", *The Art Bulletin* 63 (1981), 447-75.

Kavaler, Ethan, M., *Pieter Bruegel: Parables of Order and Enterprise* (Cambridge, 1999).

Kren, Thomas and Scott McKendrick, *Illuminating the Renaissance: The Triumph of Flemish Manuscript Painting in Europe* (Los Angeles, 2003).

Lester, G. A. *Three Late Medieval Morality Plays: Mankind, Everyman, Mundus et Infans* (London, 2002).

Malcolmson, Robert, *Popular Recreations in English Society 1700-1850* (Cambridge, 1973).

Moxey, K., "Pieter Bruegel and Popular Culture", *The Prints of Pieter Bruegel the Elder*, ed. David Freedberg (Tokyo, 1989), 42-52.

Nettekoven, Ina, *Der Meister der Apokalypsenrose der Saint Chapelle und die Pariser Buchkunst um 1500* (Turnhout, 1994).

Opie, Iona and Peter Opie, *Children's Games with Things* (Oxford, 1997).

Opie, Iona and Peter Opie, *Children's Games in Street and Playground* (Oxford, 1969).

Orenstein, Nadine, ed., *Pieter Bruegel the Elder: Drawings and Prints* (New York, 2001).

Orme, Nicholas, *Medieval Children* (New Haven, Conn., 2001).

Pächt, Otto and Jonathan J. G. Alexander, *Illuminated Manuscripts in the Bodleian Library Oxford* (Oxford, 1966).

Panofsky, Erwin, *Early Netherlandish Painting: Its Origins and Character,* 2 vols (Cambridge, Mass., 1964).

Reinsberg-Düringsfeld, Otto von, *Calendrier Belge: Fêtes religieuses et civiles, usages croyances et practiques populaires des Belges anciens et modernes*, 2 vols. (Brussels, 1891).

Royalton Kisch, M., "The King's Crown: A Popular Print for Epiphany", *Print Quarterly* 1 (1984), 43-51.

Rozler, M., "Rattles on Good Friday and Maundy Thursday in Austria", *Folklore* 47 (1936), 393-394.

Salinger, M. M., "*The Whitsun-Bride* by Pieter Brueghel the Younger", *The Metropolitan Museum of Art Bulletin* 34 (1939), 88-90.

Smeyers, Maurits and Jan Van der Stock, *Flemish Illuminated Manuscripts 1475-1550* (Ghent, 1996).

Snow, Edward, *Inside Bruegel: The Play of Images in Children's Games* (New York, 1997).

Strutt, Joseph, *The Sports and Pastimes of the People of England* (London, 1876).

Tolnay, Charles de, *Pierre Bruegel l'ancien*, 2 vols (Brussels, 1935).

Wagenberg-ter Hoeven, A. A., "The Celebration of Twelfth Night in Netherlandish Art", *Simiolus* 22 (1994), 65-96.

Wieck, Roger, *Painted Prayers: The Book of Hours in Medieval and Renaissance Art* (New York, 1997).

Willard, James F., "Occupations of the Months in Medieval Calendars", *The Bodleian Quarterly Record* 74 (1932), 33-39.

Willemsen, A., "Medieval Children's Toys in the Netherlands: Production, Sale and Trade", in *Material Culture in Medieval Europe* (Zellik, 1997), 405-411.

Willemsen, Annemarieke, *Kinder delijt: Middeleeuws speelgoed in de Nederlanden* (Nijmegen, 1998).

—. "The Game of the Month: Playful Calendars in Ghent-Bruges Books of Hours", in *Manuscripts in Transition: Recycling Manuscripts, Texts and Images*, ed. Bridgette Dekeyzer and Jan Van der Stock (Leuven, 2005), 419-430.

CONTRIBUTORS

LAURA CLEAVER is a lecturer in medieval art at The Courtauld Institute of Art. Her research concentrates on medieval manuscripts, and her interests include the relationships between art and education in the twelfth and thirteenth centuries, monastic libraries, and the uses of art in the cults of saints.

MICHAEL FOSTER is an assistant professor of medieval English literature at Korea University. His research focuses on the application of classical rhetorical theory to medieval literature and on the production and transmission of manuscripts in late-medieval England.

JOHN LANCE GRIFFITH is an associate professor of medieval literature in the Department of English at National Taipei University of Technology. His research interests in the field of medieval studies include Chaucer, romance, and modern medievalism.

LUCIANA MEINKING GUIMARÃES'S main research areas are in Welsh and Anglo-Norman hagiography and Arthurian studies. She works as an independent scholar and is currently teaching at the University of Freiburg, Germany.

TOM NICKSON is a lecturer in medieval art and architecture in the department of history of art, University of York. Tom works on the art, architecture and urbanism in medieval Spain, with a special focus on contacts between different confessional communities.

AMY ORROCK is completing a PhD on Pieter Bruegel the Elder in the department of the history of art, University of Edinburgh. Her research interests lie in representations of popular culture in the fifteenth and sixteenth centuries.

NANCY ROSS is an adjunct professor of art history at Dixie State College of Utah. Nancy works on English illuminated manuscripts from the thirteenth and fourteenth centuries, with a particular emphasis on Apocalypse manuscripts.

KAREN WARD is a PhD candidate studying medieval English cycle drama at the University of Waterloo in Ontario, Canada. Her main research interests include metatheatricality and anachronism in both medieval and contemporary productions of cycle drama.

INDEX